Ordinary
Differential Equations
using MATLAB®

SECOND EDITION

Ordinary Differential Equations
using MATLAB®

SECOND EDITION

John C. Polking
David Arnold

PRENTICE HALL, Upper Saddle River, NJ 07458

Executive Editor: George Lobell
Editorial Assistant: Gale A. Epps
Special Projects Manager: Barbara A. Murray
Production Editor: Barbara A. Till
Supplement Cover Manager: Paul Gourhan
Supplement Cover Designer: Liz Nemeth
Manufacturing Buyer: Alan Fischer

Printed in the United States of America

10 9 8 7 6 5 4 3 2 1

ISBN 0-13-011381-6

Prentice-Hall International (UK) Limited, London
Prentice-Hall of Australia Pty. Limited, Sydney
Prentice-Hall Canada, Inc., Toronto
Prentice-Hall Hispanoamericana, S.A., Mexico
Prentice-Hall of India Private Limited, New Delhi
Prentice-Hall (Singapore) Pte. Ltd.,
Prentice-Hall of Japan, Inc., Tokyo
Editora Prentice-Hall do Brazil, Ltda., Rio de Janeiro

Contents

Preface

It has been almost five years since the publication of the first edition of *Ordinary Differential Equations, Using MATLAB* *. During this time period the Mathworks has released MATLAB 5, both the professional edition as well as the *Student Edition of MATLAB 5*. In addition, there have been several significant upgrades to MATLAB 5 (the current release is MATLAB 5.3), and, of course, the accompanying toolboxes have also seen changes and improvements. Thus, it was essential that *Ordinary Differential Equations, Using Matlab* adjust to reflect the changes and advances in the MATLAB software.

All of the commands in this second edition have been thoroughly tested with MATLAB, version 5.0, and the Symbolic Toolbox, version 2.0. All commands in the manual and accompanying the routines, `dfield5` and `pplane5`, will run equally well with later releases of MATLAB 5 and the Symbolic Toolbox. If you are currently using MATLAB 4, then you will either need to upgrade your software or consider using the first edition of *Ordinary Differential Equations, Using MATLAB*. If your version of the Symbolic Toolbox is a release prior to version 2.0, but you are using MATLAB 5 or a later release, you can skip the parts in this second edition that require the use of the Symbolic Toolbox without fearing a loss of continuity. The majority of the manual (about 90-95%) relies strictly on the kernel of routines provided by MATLAB 5 and the routines provided with the manual.

Although the software has seen vast change, some things still remain the same. We still feel that there is no elementary college level mathematics course for which computer graphics is more useful to the student than the ordinary differential equations course. This manual reflects our ongoing efforts to institute a significant computer component in the ODE courses taught at Rice University and College of the Redwoods.

The manual is designed to be used by the student while working at the computer. The student should follow the manual, executing the commands as they occur in the text. Pains have been taken to make it as easy as possible for the student to work through the manual with a minimum of assistance. Of course, some assistance will inevitably be required.

Understand that the manual is meant to be a supplementary text for students in an ODE course. It is not a textbook, nor was it written as a companion for any particular textbook. In fact, this manual can accompany most ODE textbooks (we'd like to think all), and we leave the choice of text to the instructor. Although the manual is basically a book on how to use and solve ODEs with MATLAB, there are theorems stated, and there are theoretical results quoted in the manual. However, there are no proofs.

---

* MATLAB is a registered trademark of
  The MathWorks, Inc.
  For MATLAB product information, please contact
  The MathWorks, Inc.
  24 Prime Park Way
  Natick, MA 01760-1500
  Phone: 508-647-7000
  Fax: 508-647-7101
  Email: info@mathworks.com
  Web: www.mathworks.com

With a plethora of excellent ODE software packages to choose from, we have often been asked why we chose MATLAB for our students. We wanted our students to spend their time learning about ODEs and not about software. That meant that the software should be as easy as possible to use. Another consideration, possibly in conflict with the first, was that if our students were going to spend time learning software, then it would be desirable if that knowledge were useful in other venues after the course was over. These thoughts led us to look closely at the mathematical computer programs MATLAB, Maple, and Mathematica.

After spending a lot of time working with these three, we decided that MATLAB was by far the easiest to use. There were two drawbacks. First, MATLAB did not, at that time, have a symbolic computing capability, but we decided that that was not as important as the ability to solve ODEs, and to graphically display the solutions. MATLAB does this better than the other two programs. Furthermore, the Symbolic Toolbox has since become available from The MathWorks, providing an interface to the symbolic computing power of Maple. The second drawback was more serious. When teaching ODEs, it is highly desirable to have easy-to-use routines that allow students to observe the direction fields of first order ODEs, and at a more advanced level, to observe the vector fields of planar autonomous systems. MATLAB had neither of these, but we had been using MATLAB for some time, and we knew that it would not be difficult to provide these routines for our students by writing programs in MATLAB.

So we chose MATLAB. We have been pleased with the results. By all indications, our students have been pleased as well.

The emphasis in the manual is on ODEs. The features of MATLAB are explained only in so far as they apply to the study of ODEs. Given that as a guiding principle, the user of this manual will nevertheless learn a great deal about MATLAB in the process. There is no programming required in the use of the manual, unless you call

```
function xpr = xsqmt(t,x)
xpr = x^2 - t;
```

a program. We do not. There are some exercises that require what we do call programming, but in these cases the code is provided.

The first chapter is extremely elementary. It is meant for the user with almost no experience with computers.

We believe that the early introduction of computer drawn direction fields and solution trajectories is an essential ingredient in a modern course on differential equations. We like to get our students acquainted with dfield5 as early as the first day of instruction. Consequently, chapter two describes the MATLAB function dfield5, which displays the direction field of a single, user-provided ODE of first order. This function is very easy to use, and does not require any knowledge of how MATLAB works beyond knowing how to start it.

The third chapter is fundamental to almost everything that follows. It is here that the workings of MATLAB are explained. It is not necessary at this point to dive too deeply into the depths of MATLAB syntax. It is only necessary to understand how MATLAB works with matrices, and how to write function M-files of minimal complexity (like the one shown above). The graphical features of MATLAB are explored. Students learn this material easily.

Chapter four uses MATLAB to explore numerical methods of solving ODEs. Using MATLAB, it is quite easy to explore the errors actually made when the standard one-step solution methods, such as

Euler and Runge-Kutta, are used.

In chapter five we discuss advanced use of the `dfield5` routine introduced in Chapter 3. Engineering students will find this chapter particularly useful, because the chapter highlights and explains forcing functions typically encountered in engineering courses (square waves, pulse trains, etc.).

Chapter 6 now reflects the pedagogical reality that systems of ODEs are being introduced to students much earlier in modern course syllabi. This chapter now introduces elementary use of `pplane5`, a routine designed to solve planar, autonomous systems of ODEs. We deliberately put off advanced discussion of `pplane5` until Chapter 11.

Without exaggeration, chapter seven is one of the most important chapters in the manual. It is here that MATLAB's suite of ODE solvers is described, and where the student will learn how to solve essentially arbitrary systems of ODEs numerically. Learning these skills should be a goal of a modern ODE course. Given a choice, it is almost certain that most students will choose this analysis tool in later courses and job experiences.

Chapter eight introduces students to the Symbolic Toolbox. Although this interface to the symbolic computing power of Maple can perform extraordinary tasks, we concentrate solely on its performance in the solution of ODEs and systems of ODEs.

Chapter nine discusses the linear algebra required to master the use of MATLAB to solve linear systems of equations. Many ODE books assume that students know this important topic, an assumption which is often unwarranted. For that reason the discussion in chapter nine is fairly complete.

In chapter ten we apply the material learned in chapter nine to the solution of homogeneous, linear systems of ODEs with constant coefficients. The ease with which MATLAB does matrix algebra allows students to experience a wider variety of systems than they would without the use of MATLAB. In this second edition, we now feature extensive discussion of the exponential matrix and its use in the solution of linear systems.

In chapter eleven, we return to discuss advanced use of `pplane5`, the MATLAB routine first introduced in chapter six. This chapter features discussion of nullclines and non-linear, planar, autonomous systems. Here we introduce the Jacobian matrix and linearization, separatices, basins of attraction, and a set of problems designed to help students list all possible classifications of the equilibrium point of linear, planar, autonomous systems. The trace-determinant plane is used as a vehicle for classification.

The order of the chapters is close to the order of the topics in many ODE books. However, the prerequisites for the chapters are frequently minimal. Here is a complete list of the major dependencies:

$$1 \to 2$$
$$1 \to 3 \to 8$$
$$1 \to 3 \to 4 \to 7$$
$$1 \to 2 \to 3 \to 5$$
$$1 \to 3 \to 9 \to 10$$
$$1 \to 3 \to 6$$
$$1 \to 3 \to 6 \to 9 \to 10 \to 11$$

The notation $1 \to 3 \to 8$ means cover chapter one before attempting chapter 3, and cover chapter 3 before attempting chapter 8. It is clear that the material can be covered in many orders. This having been said, it must be added that toward the end of many of the chapters there is more advanced material, especially in the exercises in chapters seven, ten, and eleven.

This second edition has seen a significant increase in the number of exercises offered at the end of each chapter. We especially wanted to provide more routine problems, so that students could gradually build their confidence before tackling harder problems and what should more properly be called computer projects. We have added problems involving ODEs with forcing functions that are commonly encountered in engineering courses, and we have included a nice block of problems whose goal is to help students classify equilibrium points of linear systems. Not all of the questions are directly computer related. Many of them require the student to use the computer to make a conjecture, and then to verify their conjecture analytically. Where a problem requires the student to use aspects of MATLAB which are not explained in the text, these are explained as part of the problem.

There is special software that is needed for the manual. This includes the functions `dfield5` and `pplane5` described in chapters two, six, and eleven, together with their associated solvers, the solver routines `eul`, `rk2`, and `rk4` used in chapter four, and the square wave function `sqw` used in chapter 5. These routines do not come with either the professional or the student version of MATLAB.

The best source for the software is the webpage http://math.rice.edu/ dfield. There you will find versions which are compatible with all versions of MATLAB going back to 3.5. This is also the location where the most up to date versions of `dfield` and `pplane` will be found in the future. In addition this is the place where you can experiment with the new java based versions.

For the best compatibility with this document, you should use the versions which are meant to be used with MATLAB ver. 5 and later.

There are many people who contributed to this project. We should start with James L. Kinsey, former Dean of the Wiess School of Natural Sciences at Rice, who originally suggested this project, and who made available resources which made quick progress possible. Ken Richardson taught the new ODE course at Rice in conjunction with author Polking for the first year, and collaborated on every aspect of it. Joel Castellanos assisted with the programming at one stage, and served as a volunteer "labie."

Over the past six years, we have had conversations with many mathematicians about this manual and the associated software. We would like to mention Henry Edwards, Herman Gollwitzer, Larry Shampine, Bob Devaney, Bob Williams, Al Taylor, Marty Golubitsky, and Beverly West. To those whom we did not mention, we offer our gratitude and apologies. Special thanks go to Larry Shampine and Mark Reichelt, who allowed us to have an early view of their MATLAB ODE suite, which is partially described in Chapter seven.

The manuscript was prepared using YandY TEX, using a modified set of macros originally written by Jim Carlson. Michael Spivak was always ready with an answer to our many questions about TEX.

The students at Rice and College of the Redwoods deserve our special gratitude. They suffered through years of experimentation, complaining only when appropriate. They were very active participants in the preparation of the manual in this form.

David Arnold,
Eureka, California

John C. Polking,
Houston, Texas

March 1999

Ordinary Differential Equations
using MATLAB®
SECOND EDITION

1.   Introduction to MATLAB

MATLAB is an interactive, numerical computation program. It has powerful built-in routines for enabling a very wide variety of computations. It also has easy to use graphics commands that make the visualization of results immediately available. In some installations MATLAB will also have a Symbolic Toolbox which allows MATLAB to perform symbolic calculations as well as numerical calculations. In this chapter we will describe how MATLAB handles simple numerical expressions and mathematical formulas.

MATLAB is available on almost every computer system. Its interface is similar regardless of the system being used. In this Manual we are going to assume that the user has sufficient understanding of the computer he or she is using to start up MATLAB. At any rate we will assume that you have started up MATLAB, and that you are now faced with a window on your computer which contains the MATLAB prompt[1], >>, and a cursor waiting for you to do something. This is called the MATLAB Command Window, and it is time to begin.

Numerical Expressions

In its most elementary use, MATLAB is an extremely powerful calculator, with many built-in functions, and a very large and easily accessible memory. Let's start at the very beginning. Suppose you want to calculate a number such as $12.3(48.5 + \frac{342}{39})$. You can accomplish this using MATLAB by entering `12.3*(48.5+342/39)`. Try it. You should get the following:

```
>> 12.3*(48.5+342/39)
ans =
  704.4115
```

Notice that what you enter into MATLAB does not differ greatly from what you would write on a piece of paper. The only changes from the algebra that you use every day are the different symbols used for the algebraic operations. These are standard in the computer world, and are made necessary by the unavailability of the standard symbols on a keyboard. Here is a partial list of symbols used in MATLAB.

+	addition
−	subtraction
*	multiplication
/	right division
\	left division
^	exponentiation

While + and − have their standard meanings, * is used to indicate multiplication. You will notice that division can be indicated in two ways. The fraction $\frac{2}{3}$ can be indicated in MATLAB as either 2/3 or as 3\2. These are referred to as right division and left division, respectively.

---

[1] In the narrative that follows, readers are expected to enter the text that appears after the command prompt (>>). You must press the **Enter** or **Return** key to execute the command.

```
>> 2/3
ans =
    0.6667
>> 3\2
ans =
    0.6667
```

Exponentiation is quite different in MATLAB; it has to be, since MATLAB has no way of entering superscripts. Consequently, the power 4^3 must be entered as 4^3.

```
>> 4^3
ans =
    64
```

The order in which MATLAB performs arithmetic operations is exactly that taught in high school algebra courses. Exponentiations are done first, followed by multiplications and divisions, and finally by additions and subtractions. The standard order of precedence of arithmetic operations can be changed by inserting parentheses. For example, the result of 12.3*(48.5+342)/39 is quite different than the similar expression we computed earlier, as you will discover if you try it.

MATLAB allows the assignment of numerical values to variable names. For example, if you enter

```
>> x=3
x =
    3
```

then MATLAB will remember that x stands for 3 in subsequent computations. Therefore, computing 2.5*x will result in

```
>> 2.5*x
ans =
    7.5000
```

You can also assign names to the results of computations. For example,

```
>> y=(x+2)^3
y =
    125
```

will result in y being given the value $(3 + 2)^3 = 125$.

You will have noticed that if you do not assign a name for a computation, MATLAB will assign the default name ans to the result. This name can always be used to refer to the results of the previous computation. For example:

```
>> 2+3
ans =
    5

>> ans/5
ans =
    1
```

MATLAB has a number of preassigned variables or constants. The constant $\pi = 3.14159...$ is given the name pi.

```
>> pi
ans =
    3.1416
```

The square root of -1 is i.

```
>> sqrt(-1)
ans =
        0 + 1.0000i
```

Engineers and physicists frequently use i to represent current, so they prefer to use j for the square root of -1. MATLAB is well aware of this preference.

```
>> j
ans =
        0 + 1.0000i
```

There is no symbol for e, the base of the natural logarithms, but this can be easily computed as exp(1).

```
>> exp(1)
ans =
    2.7183
```

Mathematical Functions

There is a long list of mathematical functions that are built into MATLAB. Included are all of the functions that are standard in calculus courses.

Elementary Functions

abs(x)	The absolute value of x, i.e. $|x|$.
sqrt(x)	The square root of x, i.e. $\sqrt{x}$.
sign(x)	The signum of x, i.e. 0 if $x = 0$, -1 if $x < 0$, and $+1$ if $x > 0$.

The Trigonometric Functions

sin(x)	The sine of x, i.e. $\sin(x)$.
cos(x)	The cosine of x, i.e. $\cos(x)$.
tan(x)	The tangent of x, i.e. $\tan(x)$.
cot(x)	The cotangent of x, i.e. $\cot(x)$.
sec(x)	The secant of x, i.e. $\sec(x)$.
csc(x)	The cosecant of x, i.e. $\csc(x)$.

The Inverse Trigonometric Functions

`asin(x)`	The inverse sine of x, i.e. $\arcsin(x)$ or $\sin^{-1}(x)$.
`acos(x)`	The inverse cosine of x, i.e. $\arccos(x)$ or $\cos^{-1}(x)$.
`atan(x)`	The inverse tangent of x, i.e. $\arctan(x)$ or $\tan^{-1}(x)$.
`acot(x)`	The inverse cotangent of x, i.e. $\text{arccot}(x)$ or $\cot^{-1}(x)$.
`asec(x)`	The inverse secant of x, i.e. $\text{arcsec}(x)$ or $\sec^{-1}(x)$.
`acsc(x)`	The inverse cosecant of x, i.e. $\text{arccsc}(x)$ or $\csc^{-1}(x)$.

The Exponential and Logarithm Functions

`exp(x)`	The exponential of x, i.e. e^x.
`log(x)`	The natural logarithm of x, i.e. $\ln(x)$
`log10(x)`	The logarithm of x to base 10, i.e. $\log_{10}(x)$.

The Hyperbolic Functions

`sinh(x)`	The hyperbolic sine of x, i.e. $\sinh(x)$.
`cosh(x)`	The hyperbolic cosine of x, i.e. $\cosh(x)$.
`tanh(x)`	The hyperbolic tangent of x, i.e. $\tanh(x)$.
`coth(x)`	The hyperbolic cotangent of x, i.e. $\coth(x)$.
`sech(x)`	The hyperbolic secant of x, i.e. $\text{sech}(x)$.
`csch(x)`	The hyperbolic cosecant of x, i.e. $\text{csch}(x)$.

The Inverse Hyperbolic Functions

`asinh(x)`	The inverse hyperbolic sine of x, i.e. $\sinh^{-1}(x)$.
`acosh(x)`	The inverse hyperbolic cosine of x, i.e. $\cosh^{-1}(x)$.
`atanh(x)`	The inverse hyperbolic tangent of x, i.e. $\tanh^{-1}(x)$.
`acoth(x)`	The inverse hyperbolic cotangent of x, i.e. $\coth^{-1}(x)$.
`asech(x)`	The inverse hyperbolic secant of x, i.e. $\text{sech}^{-1}(x)$.
`acsch(x)`	The inverse hyperbolic cosecant of x, i.e. $\text{csch}^{-1}(x)$.

For a more extensive list of the functions available, see the MATLAB *User's Guide,* or MATLAB's online documentation[2]. All of these functions can be entered at the MATLAB prompt either alone or in combination. For example, to calculate $\sin(x) - \ln(\cos(x))$, where $x = 6$, we simply enter

```
>> x=6
x =
     6

>> sin(x)-log(cos(x))
ans =
   -0.2388
```

---

[2] MATLAB 5 comes with extensive online help. Typing `helpdesk` at the MATLAB prompt should open MATLAB's helpdesk in your internet browser. You can also access MATLAB's standard help files by typing `help` at the MATLAB prompt. For a list of MATLAB's elementary functions, type `help elfun` at the MATLAB prompt. It might be a good idea to bookmark MATLAB's help file at this point in your browser. You are apt to return to this destination many times in the future.

Take special notice that $\ln(\cos(x))$ is entered as `log(cos(x))`. The function `log` is MATLAB's representation of the natural logarithm function.

Output Format

Up to now we have let MATLAB repeat everything that we enter at the prompt. Sometimes this is not useful, particularly when the output is pages in length. To prevent MATLAB from echoing what we tell it, simply enter a semicolon at the end of a command. For example, enter

```
>> q=7;
```

and then ask MATLAB what it thinks q is by entering

```
>> q
q =
    7
```

If you use MATLAB to compute $\cos(\pi)$, you get

```
>> cos(pi)
ans =
   -1
```

In this case MATLAB is smart enough to realize that the answer is an integer and it displays the answer in that form. However, cos(3) is not an integer, and MATLAB gives us `-0.9900` as its value. Thus, if MATLAB is not sure that a number is an integer, it displays five significant figures in its answer. As another example, 1.57 is very close to $\pi/2$, and $\cos(\pi/2) = 0$. MATLAB gives us

```
>> cos(1.57)
ans =
    7.9633e-004
```

This is an example of MATLAB's exponential, or scientific notation. It stands for 7.9633×10^{-4}, or 0.00079633. In this case MATLAB again displays five significant digits in its answer. All of these illustrate the default format, which is called the `short` format. It is important to realize that although MATLAB only displays five significant digits in the default format, it is computing the answer to an accuracy of sixteen significant figures.

There are several other formats. We will discuss two of them. If it is necessary or desirable to have more significant digits displayed, enter `format long` at the MATLAB prompt. MATLAB will then display about sixteen significant digits. For example,

```
>> format long
>> cos(1.57)
ans =
    7.963267107332634e-004
```

There is another output format which we will find useful. If you enter `format rat`, then all numbers will be shown as rational numbers. This is called the rational format. If the numbers are actually irrational, MATLAB will find a very close rational approximation to the number.

```
>> cos(1.57)
ans =
      47/59021
```

The rational format is most useful when you are working with numbers you know to be rational.

After using a different format, you can return to the standard, short format by entering

```
>> format short
```

Complex Arithmetic

One of the nicest features of MATLAB is that it works as easily with complex numbers as it does with real numbers. The complex number $z = 2 - 3i$ is entered exactly as it is written.

```
>> z=2-3i
z =
      2.0000 - 3.0000i
```

Then if we enter $w = 3 + 5i$, we can calculate sums, products, and quotients of these numbers in exactly the same way we do for real numbers. For example,

```
>> w=3+5i;
>> z*w
ans =
      21.0000 + 1.0000i
```

and

```
>> z/w
ans =
      -0.2647 - 0.5588i
```

Any of the arithmetic functions listed earlier can be applied to complex numbers. For example,

```
>> y=sqrt(w)
y =
      2.1013 + 1.1897i
```

and

```
>> y*y
ans =
      3.0000 + 5.0000i
```

Since $y^2 = w$, it is a square root of the complex number w. The reader might try cos(w) and exp(w). In particular, the reader might wish to verify Euler's formula

$$e^{i\theta} = \cos(\theta) + i \sin(\theta)$$

6

for several values of θ, including $\theta = 2\pi, \pi, \pi/2$.

```
>> theta=pi;exp(i*theta),cos(theta)+i*sin(theta)
ans =
  -1.0000 + 0.0000i
ans =
  -1.0000 + 0.0000i
```

Note that several MATLAB commands can be placed on a single line, separated by commas, or semicolons, should you desire to suppress the output.

The ease with which MATLAB handles complex numbers has one drawback. There is at least one case where the answer is not the one we expect. Use MATLAB to calculate $(-1)^{1/3}$. Most people would expect the answer -1, but MATLAB gives us

```
>> (-1)^(1/3)
ans =
   0.5000 + 0.8660i
```

At first glance this may seem strange, but if you cube this complex number you do get -1. Consequently MATLAB is finding a *complex* cube root of -1, while we would expect a real root. The situation is even worse, since in most of the cases where this will arise in this manual, it is not the complex cube root we want. We will want the cube root of -1 to be -1.

However, this is a price we have to pay for other benefits. For MATLAB to be so flexible that it can calculate roots of arbitrary order of arbitrary complex numbers, it is necessary that it should give what seems like a strange answer for the cube root of negative numbers. In fact the same applies to any odd root of negative numbers. What we need is a way to work around the problem.

Notice that if $x < 0$, then $x = -1 \times |x|$, and we can find a negative cube root as $-1 \times |x|^{1/3}$. Here we are taking the real cube root of the positive number $|x|$, and MATLAB does that the way we want it done. But suppose the situation arises where we do not know beforehand whether x is positive or negative. What we want is:

$$x^{1/3} = \begin{cases} |x|^{1/3}, & \text{if } x > 0; \\ 0, & \text{if } x = 0; \\ -1 \times |x|^{1/3}, & \text{if } x < 0. \end{cases}$$

To write this more succinctly we use the *signum* function $\text{sgn}(x)$ (in MATLAB it is denoted by $\text{sign}(x)$). This function is defined to be

$$\text{sgn}(x) = \begin{cases} 1, & \text{if } x > 0; \\ 0, & \text{if } x = 0; \\ -1, & \text{if } x < 0. \end{cases}$$

Thus, in all cases we have $x = \text{sgn}(x)\,|x|$, and the real cube root is

$$x^{1/3} = \text{sgn}(x)\,|x|^{1/3}.$$

In MATLAB, we would enter `sign(x)*abs(x)^(1/3)`.

7

Recording Your Work

It is frequently useful to be able to record what happens in a MATLAB session. For example, in the process of preparing a homework submission, it should not be necessary to copy all of the output from the computer screen. You ought to be able to do this automatically. The MATLAB `diary` command makes this possible.

For example, suppose you are doing your first homework assignment and you want to record what you are doing in MATLAB. To do this, choose a name, perhaps hw1, for the file in which you wish to record the output. Then enter `diary hw1` at the MATLAB prompt. From this point on, everything that appears in the Command Window will also be recorded in the file `hw1`. When you want to stop recording enter `diary off`. If you want to start recording again, enter `diary on`.

The file that is created is a simple text file. It can be opened by an editor or a word processing program and edited to remove extraneous material, or to add your comments. You can also print the file to get a hard copy.

Exercises

1. Start MATLAB and begin your session with the command >> `pwd`. MATLAB will respond with the *present working directory*. If this directory is not where you wish to store your work, use MATLAB's `cd` command to change to a directory or folder of choice. For example, on a PC the command `cd c:\mywork` will change the current working directory to `c:\mywork`. You can also click the `pathtool` icon on the toolbar or enter `pathtool` at the MATLAB prompt which opens the Path Browser dialog box, after which you change the current directory interactively. Start a diary session with >> `diary hmwk1`. Read Chapter One of this manual again, but this time enter each of the commands in the narrative at the MATLAB prompt as you read. When you are finished, enter the command >> `diary off`. Open the file `hmwk1` in your favorite editor or word processor. Edit and correct any mistakes that you made. Save and print the edited file and submit the result to your instructor.

A first order ordinary differential equation has the form

$$x' = f(t, x).$$

To solve this equation we must find a function $x(t)$ such that

$$x'(t) = f(t, x(t)), \qquad \text{for all } t.$$

This means that at every point $(t, x(t))$ on the graph of x, the graph must have slope equal to $f(t, x(t))$.

We can turn this interpretation around to give a geometric view of what a differential equation is, and what it means to solve the equation. At each point (t, x), the number $f(t, x)$ represents the slope of a solution curve through this point. Imagine, if you can, a small line segment attached to each point (t, x) with slope $f(t, x)$. This collection of lines is called a *direction line field*, and it provides the geometric interpretation of a differential equation. To find a solution we must find a curve in the plane which is tangent at each point to the direction line at that point.

Admittedly, it is difficult to visualize such a direction field. This is where the MATLAB routine `dfield5` demonstrates its value[1]. Given a differential equation, it will plot the direction lines at a large number of points — enough so that the entire direction line field can be visualized by mentally interpolating between the field elements. This enables the user to get some geometric insight into the solutions of the equation.

Starting DFIELD5

To see `dfield5` in action, enter `dfield5` at the MATLAB prompt. After a short wait, a new window will appear with the label DFIELD5 Setup. Figure 2.1 shows how this window looks on a PC running Windows 95. The appearance will differ slightly depending on your computer, but the functionality will be the same on all machines.

The DFIELD5 Setup window is an example of a MATLAB *figure window*. A figure window can assume a variety of forms as will soon become apparent. In a MATLAB session there will always be one command window open on your screen and perhaps a number of figure windows as well.

You will notice that the equation $x' = x^2 - t$ is entered in the edit box entitled "The differential equation" of the DFIELD5 Setup window. There is also an edit box for the independent variable and several edit boxes are available for parameters. Note the default values in the "display window," in which the independent variable t is set to satisfy $-2 \le t \le 10$, and the dependent variable x is set to satisfy $-4 \le x \le 4$. At the bottom of the DFIELD5 Setup window there are three buttons labeled **Quit**, **Revert**, and **Proceed**.

We will describe this window in detail later, but for now click the button with the label **Proceed**. After a few seconds another window will appear, this one labeled DFIELD5 Display. An example of this window is shown in Figure 2.2.

---

[1] The MATLAB function `dfield5` is not distributed with MATLAB. To discover if it is installed properly on your computer enter `help dfield5` at the MATLAB prompt. If it is not installed, see the Preface for instructions on how to obtain it.

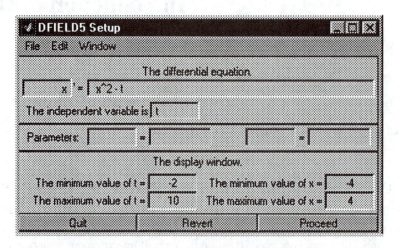

Figure 2.1. The Setup window for `dfield5`.

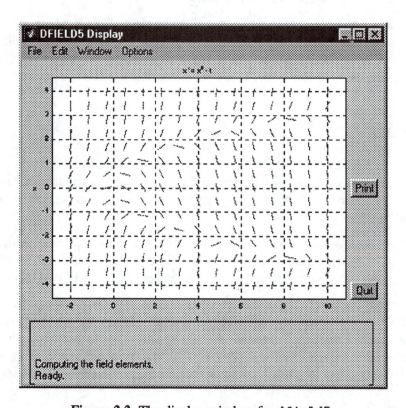

Figure 2.2. The display window for `dfield5`.

The most prominent feature of the DFIELD5 Display window is a rectangular grid labeled with the differential equation $x' = x^2 - t$ on the top, the independent variable t on the bottom, and the dependent variable x on the left. The dimensions of this rectangle are slightly larger than the rectangle specified in the DFIELD5 Setup window so as to accomodate the extra space needed by the direction field lines.

10

Inside this rectangle is a grid of points, 20 in each direction, for a total of 400 points. At each such point with coordinates (t, x) there is shown a small line segment centered at (t, x) with slope equal to $x^2 - t$.

There is a pair of buttons on the DFIELD5 Display window: **Quit** and **Print**. There are several menus: File, Edit, Window, and Options. Below the direction field there is a message window through which dfield5 will communicate with the user. Note that the last line of this window contains the word "Ready," indicating that dfield5 is ready to follow orders.

A *solution curve* of a differential equation $x' = f(t, x)$ is the graph of a function $x(t)$ which solves the differential equation. Computing and plotting a solution curve is very easy using dfield5. Choose an initial point for the solution, move the mouse to that point, and click the mouse button. The computer will compute and plot the solution through the selected point, first in the direction in which the independent variable is increasing (the "Forward" direction), and then in the opposite direction (the "Backward" direction).

After computing and plotting several solutions, the display should look something like that shown Figure 2.3.

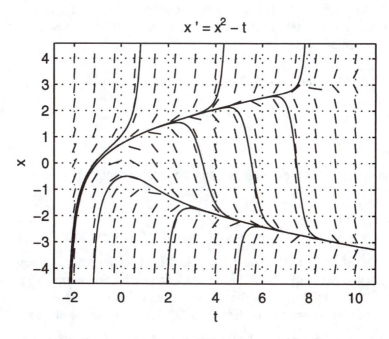

Figure 2.3. Several solutions of the ODE $x' = x^2 - t$.

Dfield5 also allows you to plot several solutions at once. Select **Options→Plot several solutions**[2] and note that the mouse cursor changes to "crosshairs" when positioned over the direction field. Select several initial conditions for your solutions by clicking the mouse button at several different locations in the direction field. When you are finished selecting initial conditions, position the mouse crosshairs

---

[2] The notation **Options→Plot several solutions** signifies that you should select "Plot several solutions" from the Options menu. We will use this notation frequently in this manual.

over the direction field and press the **Enter** or **Return** key on your keyboard. Solution trajectories will emanate from the initial conditions you selected with the mouse. Wait for the word "Ready" to appear in the message window before continuing with further computation.

You can also choose the direction of solution trajectories: Forward, Backward, or Both. Select **Options→Solution direction→Forward**, then click your mouse in the direction field and note the effect this option has on solution directions. Experiment further with **Options→Solution direction→ Back** and **Options→Solution direction→Both**.

Initial Value Problems

The differential equation $x' = x^2 - t$ has infinitely many solutions. This is suggested by the fact that you get a solution no matter where you click in the display window. Sometimes you need to find a particular solution of a differential equation, a solution that satisfies some initial condition. The differential equation with initial condition

$$x' = f(t, x), \quad x(t_0) = x_0,$$

is called an *initial value problem*. In this case, the dependent variable in the equation $x' = f(t, x)$ is x and the independent variable is t. However, you are free to choose other letters to represent the dependent and/or independent variables[3].

Example 1. *Use* `dfield5` *to find the solution of the initial value problem*

$$y' + y = 3 + \cos x, \quad y(0) = 1,$$

on the interval $0 \leq x \leq 20$.

The dependent variable in this example is y and the independent variable is x. Solve the differential equation for y'.

$$y' = -y + 3 + \cos x$$

The equation $y' = -y + 3 + \cos x$ is now in the form $y' = f(x, y)$, where $f(x, y) = -y + 3 + \cos x$.

Return to the DFIELD5 Setup window and select **Edit→Clear all**. Options on the Edit menu clear particular regions of the DFIELD5 Setup window and each of these options possesses a keyboard accelerator. Enter the left and right sides of the differential equation $y' = -y + 3 + \cos x$, the independent variable, and define the display window in the DFIELD5 Setup window as shown in Figure 2.4.

Should your data entry become hopelessly mangled, click the **Revert** button to restore the original entries. The initial value problem $y' = -y + 3 + \cos x$, $y(0) = 1$, contains no parameters, so leave the parameter fields in the DFIELD5 Setup window blank. Click the **Proceed** button to transfer the information in the DFIELD5 Setup window to the DFIELD5 Display window and start the computation of the direction field.

Choosing the initial point for the solution curve with the mouse is convenient, but sometimes it is necessary to start a solution at an exact initial condition. This is difficult to accomplish with the mouse. Instead, in the DFIELD5 Display window, select **Options→Solution direction→Both**, then **Options→ Keyboard input**. Enter the initial condition, $y(0) = 1$, as shown in Figure 2.5.

---

[3] MATLAB is case-sensitive. Thus, the variable Y is completely different from the variable y.

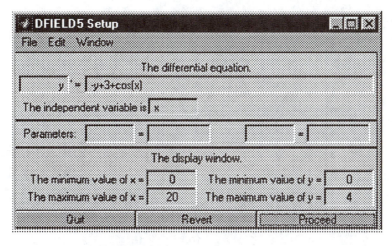

Figure 2.4. Setup window for $y' = -y + 3 + \cos x$.

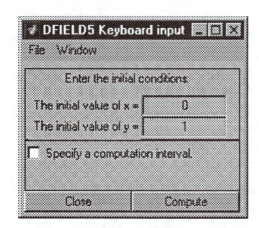

Figure 2.5. The initial condition $y(0) = 1$ starts the solution trajectory at $(0, 1)$.

Click the **Compute** button in the DFIELD5 Keyboard input window to compute the trajectory shown in Figure 2.6.

Note that you can specify a computation interval by clicking the "Specifying a computation interval" checkbox in the DFIELD5 Keyboard Input window (See Figure 2.5). Simply click the checkbox then fill in the starting and ending times of the solution interval desired. For example, start a solution trajectory with initial condition $y(0) = 0$, but set the computation interval so that $0 \leq x \leq 2\pi$. Try it!

Existence and Uniqueness

It would be comforting to know in advance that a solution of an initial value problem exists, especially if you are about to invest a lot of time and energy in an attempt to find a solution. A second (but no less important) question is uniqueness: is there only one solution? Or does the initial value problem have more than one solution? Fortunately, existence and uniqueness of solutions has been thoroughly

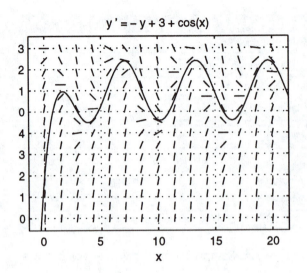

Figure 2.6. Solution of $y' + y = 3 + \cos x$, $y(0) = 1$.

examined and there is a beautiful theorem that we can use to answer these questions.

Theorem 1. *Suppose that the function $f(t, x)$ is defined in the rectangle R defined by $a \leq t \leq b$ and $c \leq x \leq d$. Suppose also that f and $\partial f / \partial x$ are both continuous in R. Then, given any point $(t_0, x_0) \in R$, there is one and only one function $x(t)$ defined for t in an interval containing t_0 such that $x(t_0) = x_0$, and $x' = f(t, x)$. Furthermore, the function $x(t)$ is defined both for $t > t_0$ and for $t < t_0$, at least until the graph of x leaves the rectangle R through one of its four edges* [4].

Example 2. *Use* `dfield5` *to sketch the solution of the initial value problem*

$$\frac{dx}{dt} = x^2, \quad x(0) = 1.$$

Set the display window so that $-2 \leq t \leq 3$ and $-4 \leq x \leq 4$.

Enter the differential equation $dx/dt = x^2$, the independent variable t, and the display window ranges $-2 \leq t \leq 3$ and $-4 \leq x \leq 4$ in the DFIELD5 Setup window. Click **Proceed** to compute the direction field. Select **Options**→ **Keyboard input** in the DFIELD5 Display window and enter the initial condition $x(0) = 1$ in the DFIELD5 Keyboard input window. If all goes well, you should produce an image similar to that in Figure 2.7.

The differential equation $dx/dt = x^2$ is in the form $x' = f(t, x)$, with $f(t, x) = x^2$. Note also that $f(t, x) = x^2$ and $\partial f / \partial x = 2x$ are continuous on the rectangle R defined by $-2 \leq t \leq 3$ and $-4 \leq x \leq 4$. Therefore, Theorem 1 states that the solution shown in Figure 2.7 is unique. Use the mouse

---

[4] The notation $\partial f / \partial x$ represents the *partial derivative of f with respect to x*. Suppose, for example, that $f(t, x) = x^2 - t$. To find $\partial f / \partial x$, think of t as a constant and differentiate with respect to x to obtain $\partial f / \partial x = 2x$. Similarly, to find $\partial f / \partial t$, think of x as a constant and differentiate with respect to t to obtain $\partial f / \partial t = -1$.

14

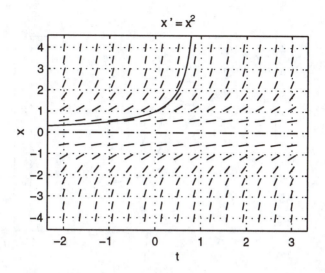

Figure 2.7. The solution of $dx/dt = x^2$, $x(0) = 1$ is unique.

to experiment. You should not be able to find any other solution trajectories that pass through the point $(0, 1)$.

Theorem 1 makes no guarantees about extending the solution in time (forward or backward). For example, does the solution in Figure 2.7 move upward and to the right forever? Or, does it reach positive infinity in a finite amount of time? This question cannot be determined by dfield5 alone. However, the differential equation $dx/dt = x^2$ is separable, yielding a family of solutions $x(t) = 1/(C-t)$, where C is an arbitrary constant. If you substitute the initial condition $x(0) = 1$ into the equation $x(t) = 1/(C-t)$, then $1 = 1/(C - 0)$, producing the constant $C = 1$ and the solution $x(t) = 1/(1 - t)$. The solution equation $x(t) = 1/(1-t)$ implies that $\lim_{t \to 1-} x(t) = +\infty$. Mathematicians like to say that the solution "blows up[5]." In this particular case, if the independent variable t represents time (in seconds), then the solution trajectory reaches positive infinity before one second of time elapses.

Example 3. *Consider the differential equation*

$$\frac{dx}{dt} = x^2 - t.$$

Sketch solutions with initial conditions $x(2) = 0$, $x(3) = 0$, and $x(4) = 0$. Determine whether or not these solution curves intersect in the display window defined by $-2 \le t \le 10$ and $-4 \le x \le 4$.

Enter the differential equation $x' = x^2 - t$, the independent variable t, and the display window ranges $-2 \le t \le 10$ and $-4 \le x \le 4$ in the DFIELD5 Setup window. Click **Proceed** to transfer this information and begin computation of the direction field in the DFIELD5 Display window. Select **Options→Keyboard input** in the DFIELD5 Display window and compute solutions for each of the initial conditions $x(2) = 0$, $x(3) = 0$, and $x(4) = 0$. If all goes well, you should produce an image similar to that in Figure 2.8.

---

[5] The graph of the solution reaches infinity (or negative infinity) in a finite time period.

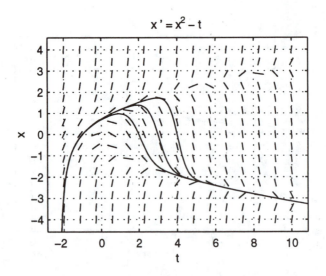

Figure 2.8. Do the trajectories intersect?

When you compare the ODE $x' = x^2 - t$ with general form $x' = f(t, x)$, you see that $f(t, x) = x^2 - t$ is continuous on the display window defined by $-2 \leq t \leq 10$ and $-4 \leq x \leq 4$. Moreover, $\partial f / \partial x = 2x$ is also continuous on this display window. In Figure 2.8, it appears that the solution trajectories merge into one trajectory near the point $(8, -2.7)$ (or perhaps even sooner). However, Theorem 1 guarantees that solutions cannot cross or meet in the display window of Figure 2.8.

Let's do some analysis with the zoom tools in dfield5. On a PC[6], you would select **Edit→Zoom in** in the DFIELD5 Display window, then use the left mouse button and single-click in the DFIELD5 Display window near the point $(8, -2.7)$. Additional "zooms" require that you revisit **Edit→Zoom in** before left-clicking the mouse button to zoom. There is a faster way to zoom in that is platform dependent. For example, on a PC or UNIX box, click the right mouse button at the zoom point (or control-click the left mouse button at the zoom point). On a Macintosh, option-click the mouse button at the zoom point. After performing numerous zooms (results may vary on your machine), each time clicking the right mouse button on the trajectory near $t = 8$, some separation in the trajectories begins to occur, as shown in Figure 2.9. Without Theorem 1, we might have mistakenly assumed that the trajectories merged into one trajectory.

Qualitative Analysis

Suppose that you model a population with a differential equation. If you want to use your model to predict the exact population in three years, then you will need to do one of two things: (1) find the solution using analysis, or (2) use a numerical routine to find the solution. However, if your only interest is what happens to the population after a long period of time, it may not be necessary to find an analytical or numerical solution. A qualitative approach might be more appropriate.

---

[6] Mouse actions are platform dependent in dfield5. See the front and back covers of this manual for a summary of mouse actions on various platforms.

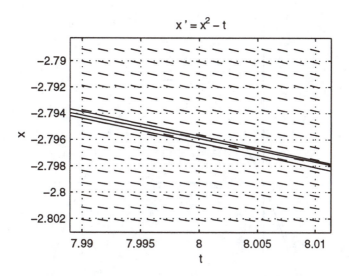

Figure 2.9. The trajectories don't merge or cross.

Example 4. *Consider the logistic population model*

$$\frac{dP}{dt} = k\left(1 - \frac{P}{N}\right)P. \tag{2.1}$$

Let $P(t)$ represent the population at time t. Let $k = 0.75$ and $N = 10$ and suppose that the initial population at time zero is given by $P(0) = 2$, where the population is measured in millions of people. What will happen to this population over a long period of time?

If you plot the right hand side of equation (2.1) versus P (i.e., plot $k(1 - P/N)P$ versus P), the result is the inverted parabola seen in Figure 2.10. Set $k(1 - P/N)P$ equal to zero to find that the graph crosses the P-axis in Figure 2.10 at $P = 0$ and $P = N$.

It is easily verified that $P(t) = N$ is a solution of $dP/dt = k(1 - P/N)P$ by substituting $P(t) = N$ into each side of the differential equation and simplifying. Similarly, the solution $P(t) = 0$ is easily seen to satisfy the differential equation.

Although the solutions $P(t) = 0$ and $P(t) = N$ might be considered "trivial" since they are constant functions, they are by no means trivial in their importance. In fact, the solutions $P(t) = 0$ and $P(t) = N$ are called *equilibrium solutions*. For example, if $P(t) = N$, then the growth rate dP/dt of the population is zero and the population remains at $P(t) = N$ forever. Similarly, if $P(t) = 0$, the growth rate dP/dt equals zero and the population remains at $P(t) = 0$ for all time.

When the graph of $k(1 - P/N)P$ (which equals to dP/dt) falls below the P-axis in Figure 2.10, then $dP/dt < 0$ and the first derivative test implies that $P(t)$ is a decreasing function of t. On the other hand, when the graph of $k(1 - P/N)P$ rises above the P-axis in Figure 2.10, then $dP/dt > 0$ and $P(t)$ is an increasing function of t. These facts are summarized on the *phase line* below the graph in Figure 2.10. The information on the phase line indicates that a population beginning between 0 and N million people has to increase to the equilibrium value of N million people.

The dependent variable of the differential equation $dP/dt = k(1 - P/N)P$ is P and the independent variable is t; k and N are called *parameters*. Return to the DFIELD5 Setup window and enter the

17

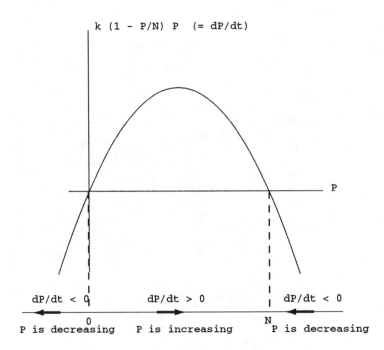

Figure 2.10. The plot of $k(1 - P/N)P$ versus P.

differential equation, the independent variable t, and the parameters $k = 0.75$ and $N = 10$ as shown in Figure 2.11. Set the display window ranges to $0 \le t \le 10$ and $-4 \le P \le 15$ (See Figure 2.11). Click the **Proceed** button to transfer the information in the DFIELD5 Setup window to the DFIELD5 Display window and start the computation of the direction field.

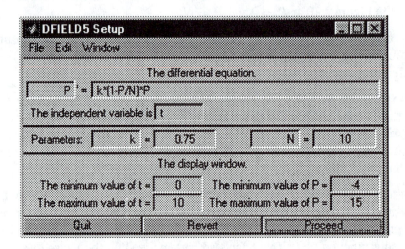

Figure 2.11. Setup window for $dP/dt = k(1 - P/N)P$.

Select **Options**→**Keyboard input** in the DFIELD5 Display window and start solution trajectories with initial conditions $(0, 0)$ and $(0, 10)$. In Figure 2.12, note that these equilibrium solutions are horizontal lines. Select **Options→Solution direction→Forward** and **Options→Show the phase line** in the DFIELD5 Display window. `Dfield5` aligns the phase line from Figure 2.10 in a vertical direction at the left edge of the direction field in the DFIELD5 Display window. Select **Options→Keyboard input** to begin the solution with initial condition $P(0) = 2$ and note the action of the animated point on the phase line. As the solution trajectory in the direction field approaches the horizontal line $P = 10$, the point on the phase line approaches equilibrium point $P = 10$ on the phase line, as shown in Figure 2.12. It would appear that a population with initial conditions and parameters described in the original problem statement will have to approach 10 million people with the passage of time.

Experiment with some other initial conditions. Note that solutions beginning a little above or a little below the equilibrium solution $P = 10$ tend to move back toward this equilibrium solution with the passage of time. This is why the solution $P = 10$ is called a *stable* equilibrium solution. However, solutions beginning a little above or a little below the equilibrium solution $P = 0$ tend to move away from this equilibrium solution with the passage of time. The solution $P = 0$ is called an *unstable* equilibrium solution. You can review the results of our experiments in Figure 2.12.

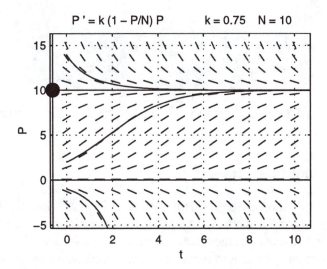

Figure 2.12. Note the vertical phase line at the left of the window.

Zooming and Stopping

It is rare that a solution takes an inordinate amount of time to draw, but should this occur you can halt computation by clicking the **Stop** button in the DFIELD5 Display window. You may have to click the **Stop** button twice: once for the forward direction, a second time to stop the computation in the backward direction. For example, enter the equation `y'=exp(2*t)*cos(y)` in the DFIELD5 Setup window, set the independent variable as `t`, then set the display window so that $-1 \le t \le 6$ and $0 \le y \le 3$. Select **Options→Solution direction→Both** and use the Keyboard input window to start a solution with initial condition $y(0) = 0$. Note that the forward solution begins to stall. Click the **Stop** button to halt the forward solution and note that the backward solution completes rather quickly.

You can use the mouse and/or menu selections to "zoom in" or "zoom back" in the DFIELD5 Display window. PC users can drag a "zoom box" around the solution of y'=exp(2*t)*cos(y) by depressing the right mouse button, then dragging the mouse. Once the zoom box is drawn around the area of interest, release the mouse button and the contents of the zoom box will be magnified to full size of the display window.

Dfield5 allows you to revisit any of your zoom windows. Select **Edit→Zoom back** in the DFIELD5 Display window. This will open the DFIELD5 Zoom back dialog box pictured in Figure 2.13. Select the zoom window you wish to revisit and click the **Zoom** button.

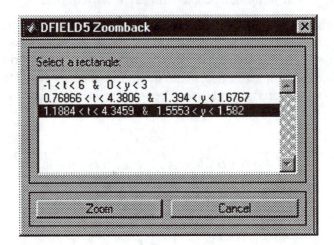

Figure 2.13. Select a zoom window and click the **Zoom** button.

Using MATLAB While DFIELD5 is Open

You can use MATLAB commands to plot to the DFIELD5 Display window; or, you can open another figure window by typing figure at the MATLAB prompt. Future plotting commands will be directed to this window, as long as you do not click the mouse while the cursor is in another window. Remember, all graphics commands will usually be sent to the *active* figure window, which is always the most recently visited window.

Example 5. *The following data represent the temperature T of a potato after t minutes in an oven. Use Newton's Law of Cooling and* dfield5 *to fit a curve to the data.*

t (min)	0	1	2	3	4	5	6	7	8	9	10
T (°F)	60	213	306	362	397	417	430	438	442	445	447

Newton determined that the rate at which an object warms (or cools) is proportional to the difference between the temperature of the object and its surrounding medium. Consequently,

$$\frac{dT}{dt} = k(A - T), \quad T(0) = 60,$$

where T is the temperature of the potato in degrees Fahrenheit, t is the time in minutes, A is the temperature of the oven, and k is a proportionality constant. Enter the information shown in Figure 2.14 in the DFIELD5 Setup window.

20

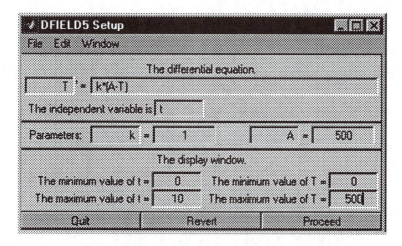

Figure 2.14. Setting up $dT/dt = k(A - T)$.

Note that we have made some wild guesses for the parameters k and A. Click on the **Proceed** button to process this information and compute the direction field in the DFIELD5 Display window. This also makes the DFIELD5 Display window the current figure window.

Enter the following commands in the MATLAB command window at the MATLAB prompt:

```
>> t=[0,1,2,3,4,5,6,7,8,9,10];
>> T=[60,213,306,362,397,417,430,438,442,445,447];
>> plot(t,T,'o')
```

Finally, select **Options→Keyboard input** and plot the trajectory with initial condition $T(0) = 60$. Experiment with the parameters k and A and the DFIELD5 Keyboard input dialog box until you find a solution that passes through each of the data points. *Hint: The parameters $k = 0.5$ and $A = 450$ produced the image in Figure 2.15.*

Changing the Size and Appearance of the Display Window

Some people prefer to use a *vector field* rather than a direction field in the DFIELD5 Display window. In a vector field, a vector is attached to each point instead of the line segment used in a direction field. The vector has its base at the point in question, its direction is the slope, and the length of the vector reflects the magnitude of the derivative.

To change the direction field in the DFIELD5 Display window to a vector field, select **Options→Windows settings** from the Options menu in the DFIELD5 Display window. This will open the DFIELD5 Windows settings dialog box (see Figure 2.16).

Note the three radio buttons that allow you to choose between a line field, a vector field, or no field at all. Select one of these and then click the **Change settings** button to note the affect on the direction field. Should you select vectors as your option, note that the length of each arrow reflects the *relative magnitude* of the derivative at that point.

There is an edit box in the DFIELD5 Window settings dialog that allows the user to choose the

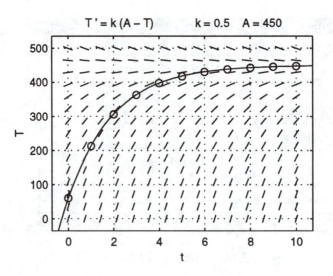

Figure 2.15. Plotting in the DFIELD5 Display window.

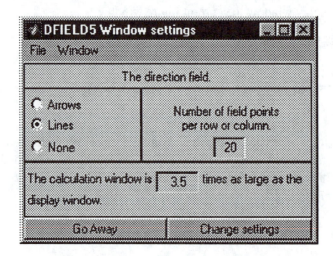

Figure 2.16. DFIELD5 Window settings.

number of field points displayed. The default is 20 points in each row and in each column. Change this entry to 10, hit **Enter**, then click the **Change settings** button to note the affect on the direction field.

The design of `dfield5` includes the definition of two windows: the DFIELD5 Display window and the *calculation window*. When you start `dfield5`, the calculation window is 3.5 times as large as the display window in each dimension. The computation of a solution will stop only when the solution curve leaves the calculation window. This allows some room for zooming to a larger display window without having incomplete solution curves. It also allows for some reentrant solution curves, i.e. those which leave the display window and later return to it.

The third item in the DFIELD5 Window settings dialog box controls the relative size of the calculation window. It can be given any value greater than or equal to 1. The smaller this number, the faster `dfield5` will compute solutions, and the more likely that reentrant solutions will be lost. If you have a slower computer, and if you are not going to be zooming to a larger display window, it is not completely unreasonable to set this value to 1, but you have to realize that with this choice all reentrant solutions will be lost. A better choice would be 2 or 2.5. The default value of 3.5 seems to meet most needs.

Personalizing the Display Window

Sometimes when you are preparing a display window for printing, you plot a solution curve you wish were not there. There are two methods which allow you to erase items in the DFIELD5 Display window. **Edit→Erase all solutions** is self explanatory. **Edit→Delete a graphics object** is much more flexible. It will allow you to delete individual solution curves. Simply select **Edit→Delete a graphics object**, then click the mouse on the solution curve you wish to delete. You can use **Edit→Delete a graphics object** to delete any graphics object in the DFIELD5 Display window, including text.

There are three text elements which are part of the display window. These are the `title` at the top, the `xlabel` at the bottom, and the `ylabel` at the left. These are given default values by `dfield5` using the information entered into the DFIELD5 Setup window, but they can be changed at any time. Should you want the title and axes labels to reflect the content of your project, try something like this.

```
>> xlabel('Time (years)')
>> ylabel('Frank''s cell count')
>> title('Petri Dish History')
```

Notice that each of the three commands takes as a parameter a text string which is contained between two single quotes. You can use any text string you think is appropriate.

There is a special problem with the `ylabel` string in this example. You might think that it should read `'Frank's cell count'`. The problem is that the prime `'` is being used both to denote an apostrophe and to indicate the beginning or end of the text string. MATLAB needs a different way to designate a prime internal to a text string, and it uses the double prime `''` to do just this.

It is also possible to add text at arbitrary points in the DFIELD5 Display window. Select **Edit→ Enter text on the Display Window**, enter the desired text in the Text entry dialog box, then click the **OK** button. Use the mouse to click at the lower left point of the position in the figure window where you want the text to appear. It can easily happen that your placement of the text does not please you. If so, remove the text using **Edit→Delete a graphics object**, then try again.

Printing, Quitting, and Using Clipboards

The easiest way to print the display window is to click the **Print** button in the DFIELD5 Display window.

If you have previous experience using MATLAB, you realize that the display window, or more generally, the current figure window, can be printed by entering `print` at the MATLAB prompt in the command window. You can also use the print command at the command line to save the contents of a figure window to a graphics file. For example,

```
>> print -deps junk.eps
```

will save the DFIELD5 Display window as an encapsulated postscript file in the current directory. The command

```
>> print -deps -noui junk.eps
```

will also save DFIELD5 Display window, but without the graphical user interface objects such as buttons and message windows. You might want to click on File on the DFIELD5 Display menu and select Page Position. This will open a dialog box that will allow the resizing of the figure window (manually or with a mouse). After you have resized the page position, retry the command `print -deps -noui junk.eps`. For a full set of options on MATLAB's `print` command, type `help print` at the MATLAB prompt.

In the Macintosh and PC-Windows version of MATLAB, the contents of the display window can be copied into a clipboard, and from there into other documents in the standard ways. For example, if you are using a PC with the Windows 95 operating system, Alt+PrintScrn (Holding down the Alt key while pressing PrintScrn) will copy the current figure window to the clipboard. Consult your computer's operating manual for details on copying and pasting images to and from your clipboard.

Always wait until the word "Ready" appears in the `dfield5` message window before you try to do anything else with `dfield5` or MATLAB. When you want to quit `dfield5`, the best way is to use the **Quit** buttons found on the DFIELD5 Setup or on the DFIELD5 Display windows. Either of these will close all of the `dfield5` windows in an orderly manner, and it will delete the temporary files that `dfield5` creates in order to do its business.

Exercises

For the differential equations in problems 1–3, perform each of the following tasks.

i) Print out the direction field for the differential equation with the display window defined by $t \in [-5, 5]$ and $y \in [-5, 5]$. You might consider increasing the number of field points to 25 in the DFIELD5 Window settings dialog box. On this printout, sketch with a pencil as best you can the solution curves through the initial points $(t_0, y_0) = (0, 0)$, $(-2, 0)$, $(-3, 0)$, $(0, 1)$, and $(4, 0)$. Remember that the solution curves must be tangent to the direction lines at each point.

ii) Use `dfield5` to plot the same solution curves to check your accuracy. Turn in both versions.

1. $y' = y^2 - t^2$
2. $y' = 2ty/(1 + y^2)$
3. $y' = y(2 + y)(2 - y)$
4. Use `dfield5` to plot a few solution curves to the equation $x' + x \sin(t) = \cos(t)$. Use the display window defined by $x \in (-10, 10)$ and $t \in (-10, 10)$.
5. Use `dfield5` to plot the solution curves for the equation $x' = 1 - t^2 + \sin(tx)$ with initial values $x = -3, -2, -1, 0, 1, 2, 3$ at $t = 0$. Find a good display window by experimentation.
6. Consider the differential equation

$$y' + 4y = 8.$$

   a) Use `dfield5` to plot a few solutions with different initial points on the display window bounded by $-5 \le t \le 5$ and $-1 \le y \le 5$. In particular, plot the solution curve with initial condition $y(1) = 2$ (use **Options**→**Keyboard input**). Print out the Figure Window and turn it in as part of this assignment.

   b) What do you conjecture is the limiting behavior of the solutions of this differential equation as $t \to \infty$?

   c) Find the general analytic solution to this equation.

   d) Verify the conjecture you made in part b), or if you no longer believe it, make a new conjecture and verify that.

7. Consider the differential equation
$$(1 + t^2)y' + 4ty = t.$$

    a) Use `dfield5` to calculate and plot a few solutions with different initial points. (Use the display window defined by $t \in [-5, 5]$ and $y \in [-5, 5]$.) In particular, plot the solution curve with initial condition $y(1) = 1/4$ (use **Options→Keyboard input**). Print out the Figure Window and turn it in as part of this assignment.

    b) What do you conjecture is the limiting behavior of the solutions of this differential equation as $t \to \infty$?

    c) Find the general analytic solution to this equation.

    d) Verify the conjecture you made in part b), or if you no longer believe it, make a new conjecture and verify that.

For Problems 8–11 we will consider a certain lake which has a volume of $V = 100$ km^3. It is fed by a river at a rate of r_i km^3/year, and there is another river which is fed by the lake at a rate which keeps the volume of the lake constant. In addition, there is a factory on the lake which introduces a pollutant into the lake at the rate of p km^3/year. This means that the rate of flow from the lake into the outlet river is $(p + r_i)$ km^3/year. Let $x(t)$ denote the volume of the pollutant in the lake at time t, and let $c(t) = x(t)/V$ denote the concentration of the pollutant.

8. Show that, under the assumption of immediate and perfect mixing of the pollutant into the lake water, the concentration satisfies the differential equation $c' + ((p + r_i)/V)c = p/V$.

9. Suppose that $r_i = 50$, and $p = 2$.

    a) Suppose that the factory starts operating at time $t = 0$, so that the initial concentration is 0. Use `dfield5` to plot the solution.

    b) It has been determined that a concentration of over 2% is hazardous for the fish in the lake. Approximately how long will it take until this concentration is reached? You can "zoom in" on the `dfield5` plot to enable a more accurate estimate.   1.413 yrs.

    c) What is the limiting concentration? About how long does it take for the concentration to reach a concentration of 3.5%?   .0383 ; 4.64 yrs.

10. Suppose the factory stops operating at time $t = 0$, and that the concentration was 3.5% at that time. Approximately how long will it take before the concentration falls below 2%, and the lake is no longer hazardous for fish? Notice that $p = 0$ for this exercise.   1.13

11. Rivers do not flow at the same rate the year around. They tend to be full in the Spring when the snow melts, and to flow more slowly in the Fall. To take this into account, suppose the flow of our river is

$$r_i = 50 + 20\cos(2\pi(t - 1/3)).$$

Our river flows at its maximum rate one-third into the year, i.e., around the first of April, and at its minimum around the first of October.

    a) Setting $p = 2$, and using this flow rate, use `dfield5` to plot the concentration for several choices of initial concentration between 0% and 4%. (You might have to reduce the relative error tolerance in **Options–>Solver settings**, perhaps to 5×10^{-12}, or `5e-12`.) How would you describe the behavior of the concentration for large values of time?

    b) It might be expected that after settling into a steady state, the concentration would be greatest when the flow was smallest, i.e., around the first of October. At what time of year does it actually occur?

12. Use `dfield5` to plot several solutions to the equation $z' = (z - t)^{5/3}$. (**Hint:** Notice that when $z < t$, $z' < 0$, so the direction field should point down, and the solution curves should be decreasing. You might have difficulty getting the direction field and the solutions to look like that. If so read the section in Chapter 1 on complex arithmetic, especially the last couple of paragraphs.)

A differential equation of the form $dx/dt = f(x)$, whose right-hand side does not explicitly depend on the independent variable t, is called an *autonomous* differential equation. For example, the logistic model in Example 4 was autonomous. For the autonomous differential equations in Problems 13–17, perform each of the following tasks. Note that the first three tasks are to be performed without the aid of technology.

    i) Set the right-hand side of the differential equation equal to zero and solve for the equilibrium points.

ii) Plot the graph of the right-hand side of each autonomous differential equation versus x, as in Figure 2.10. Draw the phase line below the graph and indicate where x is increasing or decreasing, as was done in Figure 2.10.

iii) Use the information in parts (i) and (ii) to draw sample solutions in the xt plane. Be sure to include the equilibrium solutions.

iv) Check your results with dfield5. Again, be sure to include the equilibrium solutions.

v) If x_0 is an equilibrium point, i.e., if $f(x_0) = 0$, then $x(t) = x_0$ is an equilibrium solution. It can be shown that if $f'(x_0) < 0$, then every solution curve that has an initial value near x_0 converges to x_0 as $t \to \infty$. In this case x_0 is called a *stable* equilibrium point. If $f'(x_0) > 0$, then every solution curve that has an initial value near x_0 diverges away from x_0 as $t \to \infty$, and x_0 is called an *unstable* equilibrium point. If $f'(x_0) = 0$, no conclusion can be drawn about the behavior of solution curves. In this case the equilibrium point may fail to be either stable or unstable. Apply this test to each of the equilibrium points.

13. $x' = \cos(\pi x)$, $x \in [-3, 3]$.

14. $x' = x(x - 2)$, $-\infty < x < \infty$.

15. $x' = x(x - 2)^2$, $-\infty < x < \infty$.

16. $x' = x(x - 2)^3$, $-\infty < x < \infty$.

17. $x' = x(1 + e^{-x} - x^2)$, $-1 \le x \le 2$. In this case you will not be able to solve explicitly for all of the equilibrium points. Instead, turn the problem around. Use dfield5 to plot some solutions, and from that information calculate approximately where the equilibrium points are, and determine the type of each. Check your estimate with this code:

```
f=inline('x*(1+exp(-x)-x^2)')
z=fzero(f,1)
f(z)
```

The logistic equation is

$$P' = \alpha P \left(1 - \frac{P}{N}\right).$$

The quantities α and N are the parameters of the equation. Usually the parameters are constants, and in that case the kind of analysis carried out in Problems 13–17 shows that for any solution $P(t)$ which has a positive initial value we have $P(t) \to N$ as $t \to \infty$. For this reason N is called the *carrying capacity* of the system.

There are some models in which the carrying capacity is not a constant, but depends on time. In cases like this can we say anything about the relationship between the long term behavior of the solutions and the carrying capacity? For the carrying capacities in Problems 18–21 you are to examine the long term behavior of solutions, especially in comparison to the carrying capacity. In particular:

a) Use dfield5 to plot several solutions to the equation. (It is up to you to find a display window that is appropriate to the problem at hand.)

b) Based on the plot done in part a), describe the asymptotic behavior of the solutions to the equation. In particular, compare this asymptotic behavior to the asymptotic behavior of N. It might be helpful to plot N on Display Window produced in part a). In the first two cases the solutions will be asymptotic to a constant. In the other two the solutions will be asymptotic to a function. You are not expected to find that function explicitly, but you should be able to describe it qualitatively.

c) It is possible to find the solutions to these equations explicitly (except, perhaps, for the evaluation of an integral). Find these solutions. (**Hint:** Look up Bernoulli's equation in your textbook.)

Consider the equation in the following four cases:

18. $N(t) = 1$, $\alpha = 1$. This case is the standard logistic equation. Consequently, whatever the initial population, we expect that $P(t) \to N$ as $t \to \infty$. This case is here for comparison with the other three.

26

19. $N(t) = 1 - \frac{1}{2}e^{-t}$, $\alpha = 1$. In this case $N(t)$ is monotone increasing, and $N(t)$ is asymptotic to 1. This might model a situation of a human population where, due to technological improvement, the availability of resources is increasing with time, although ultimately limited.

20. $N(t) = 1 + t$, $\alpha = 1$. Again $N(t)$ is monotone increasing, but this time it is unbounded, although of a very simple nature. This might model a situation of a human population where, due to technological improvement, the availability of resources is steadily increasing with time, and therefore the effects of competition are becoming less severe. Note: The latest version of dfield5 supports the entry of expressions in the parameter windows. After entering the parameter N in the DFIELD5 Setup window, set N equal to 1+t.

21. $N(t) = 1 - \frac{1}{2}\cos(2\pi t)$, $\alpha = 1$. This is perhaps the most interesting case. Here the carrying capacity is periodic in time with period 1, which might be considered to be one year. This might model a population of insects or small animals that are affected by the seasons. You will notice that the asymptotic behavior as $t \to \infty$ reflects the behavior of N. The solution does not tend to a constant, but nevertheless all solutions have the same asymptotic behavior for large values of t. In particular, you should take notice of the location of the maxima of N and of P. You can use the "zoom in" option to get a better picture of this. *Note: It is interesting to superimpose the plot of the carrying capacity $N(t) = 1 - \frac{1}{2}\cos(2\pi t)$ on the solution in the DFIELD5 Display window. Try the following at the MATLAB prompt:* t=linspace(a,b), *where* a *and* b *are the bounds for* t *in the DFIELD5 Display window. Then follow with* N=1-1/2*cos(2*pi*t) *and* plot(t,N).

22. Despite the seeming generality of the uniqueness theorem, there are initial value problems which have more than one solution. Consider the differential equation $y' = \sqrt{|y|}$. Notice that $y(t) \equiv 0$ is a solution with the initial condition $y(0) = 0$. (Of course by $\sqrt{|y|}$ we mean the **nonnegative** square root.)

   a) This equation is separable. Use this to find a solution to the equation with the initial value $y(t_0) = 0$ assuming that $y \geq 0$. You should get the answer $y(t) = (t - t_0)^2/4$. Notice, however, that this is a solution only for $t \geq t_0$. Why?

   b) Show that the function

   $$y(t) = \begin{cases} 0, & \text{if } t < t_0; \\ (t - t_0)^2/4, & \text{if } t \geq t_0; \end{cases}$$

   is continuous, has a continuous first derivative, and satisfies the differential equation $y' = \sqrt{|y|}$.

   c) For any $t_0 \geq 0$ the function defined in part b) satisfies the initial condition $y(0) = 0$. Why doesn't this violate the uniqueness part of the theorem?

   d) Find another solution to the initial value problem in a) by assuming that $y \leq 0$.

   e) You might be curious (as were the authors) about what dfield5 will do with this equation. Find out. Use the rectangle defined by $-1 \leq t \leq 1$ and $-1 \leq y \leq 1$ and plot the solution of $y' = \sqrt{|y|}$ with initial value $y(0) = 0$. Also, plot the solution for $y(0) = 10^{-50}$ (the MATLAB notation for 10^{-50} is 1e-50). Plot a few other solutions as well. Do you see evidence of the non-uniqueness observed in part c)?

An important aspect of differential equations is the dependence of solutions on initial conditions. There are two points to be made. First, we have a theorem which says that the solutions are continuous with respect to the initial conditions. More precisely,

Theorem. *Suppose that the function $f(t, x)$ is defined in the rectangle R defined by $a \leq t \leq b$ and $c \leq x \leq d$. Suppose also that f and $\frac{\partial f}{\partial x}$ are both continuous in R, and that*

$$\left|\frac{\partial f}{\partial x}\right| \leq L \quad \text{for all } (t, x) \in R.$$

If (t_0, x_0) and (t_0, y_0) are both in R, and if

$$\begin{array}{ccc} x' = f(t, x) & & y' = f(t, y) \\ & \text{and} & \\ x(t_0) = x_0 & & y(t_0) = y_0 \end{array}$$

27

then for $t > t_0$

$$|x(t) - y(t)| \le e^{L(t-t_0)}|x_0 - y_0|$$

as long as both solution curves remain in R.

Roughly, the theorem says that if we have initial values that are sufficiently close to each other, the solutions will remain close, at least if we restrict our view to the rectangle R. Since it is easy to make measurement mistakes, and thereby get initial values off by a little, this is reassuring.

For the second point, we notice that although the dependence on the initial condition is continuous, the term $e^{L(t-t_0)}$ allows the solutions to get exponentially far apart as the interval between t and t_0 increases. That is, the solutions can still be extremely sensitive to the initial conditions, especially over long t intervals.

23. Consider the differential equation $x' = x(1 - x^2)$.

   a) Verify that $x(t) \equiv 0$ is the solution with initial value $x(0) = 0$.

   b) Use **dfield5** to find approximately how close the initial value y_0 must be to 0 so that the solution $y(t)$ of our equation with that initial value satisfies $y(t) \le 0.1$ for $0 \le t \le t_f$, with $t_f = 2$. You can use the display window $0 \le t \le 2$, and $0 \le x \le 0.1$, and experiment with initial values in the **Options→ Keyboard input** window, until you get close enough. Do not try to be too precise. Two significant figures is sufficient.

   c) As the length of the t interval is increased, how close must y_0 be to 0 in order to insure the same accuracy? To find out, repeat part b) with $t_f = 4, 6, 8$, and 10.

It is clear from the results of the last problem that the solutions can be extremely sensitive to changes in the initial conditions. This sensitivity allows chaos to occur in deterministic systems, which is the subject of much current research.

One way to experience sensitivity to changes in the initial conditions at first hand is to try a little "target practice." For the ODEs in Problems 24–27, use **dfield5** to find approximately the value of x_0 such that the solution $x(t)$ to the initial value problem with initial condition $x(0) = x_0$ satisfies $x(t_1) = x_1$. You should use the Keyboard input window to initiate the solution. After an unsuccessful attempt try again with another initial condition. The Uniqueness Theorem should help you limit your choices. If you make sure that the Display Window is the current figure (by clicking on it), and execute **plot(t1,x1,'or')** at the command line, you will have a nice target to shoot at.

You will find that hitting the target gets more difficult in each of these problems.

24. $x' = x - \sin(x)$, $t_1 = 5$, $x_1 = 2$.

25. $x' = x^2 - t$, $t_1 = 4$, $x_1 = 0$.

26. $x' = x(1 - x^2)$, $t_1 = 5$, $x_1 = 0.5$.

27. $x' = x\sin(x) + t$, $t_1 = 5$, $x_1 = 0$.

3. Vectors, Matrices, and Array Operations

A powerful feature of MATLAB is that every numerical quantity is considered to be a complex matrix![1] For those of you who do not already know, a *matrix* is a rectangular array of numbers. For example,

$$A = \begin{bmatrix} 1 & \pi & \sqrt{-1} \\ \sqrt{2} & 4 & 0 \end{bmatrix}$$

is a matrix with 2 rows and 3 columns.

Matrices in MATLAB

If you want to enter the matrix A into MATLAB proceed as follows:

```
>> A=[1,pi,sqrt(-1);sqrt(2),4,0]
A =
   1.0000              3.1416              0 + 1.0000i
   1.4142              4.0000              0
```

Note that commas are used to separate the individual elements in a row; semicolons are used to separate the rows of the matrix. You can also use *spaces* to separate (delimit) the entries in a row.

```
>> A=[1 pi sqrt(-1);sqrt(2) 4 0]
A =
   1.0000              3.1416              0 + 1.0000i
   1.4142              4.0000              0
```

The *size* of a matrix is the number of rows and columns. For example,

```
>> size(A)
ans =
     2     3
```

verifies that A has 2 rows and 3 columns. Two matrices are said to have the same size if they have the same number of rows and the same number of columns.

Even single numbers in MATLAB are matrices.

```
>> a=5;
>> size(a)
ans =
     1     1
```

MATLAB thinks that 5, or any other number, is a matrix with one row and one column.

---

[1] A matrix whose entries are complex numbers.

Addition, Subtraction, and Scalar Multiplication

If A and B are matrices of the same size, then they can be added together. For example,

```
>> A=[1 2;3 4],B=[5,6;7,8],C=A+B
A =
      1     2
      3     4
B =
      5     6
      7     8
C =
      6     8
     10    12
```

You will notice that each element of the matrix C is sum of the corresponding elements in the matrices A and B. The same is true for the difference of two matrices. Try C-A, and see what you get (you should get B).

You can also multiply a matrix by a scalar.

```
>> A=ones(2,2),C=-5*A
A =
      1     1
      1     1
C =
     -5    -5
     -5    -5
```

Note that the matrix C is formed by multiplying each entry of the matrix A by -5.

Vectors in Matlab

A *vector* is a list of numbers.[2] It can be a vertical list, in which case it is called a *column vector*, or it can be a horizontal list, in which case it is called a *row vector*. Thus, vectors are special cases of matrices.

For example, we can define a column vector as follows:

---

[2] The word *vector* is one of the most overused terms in mathematics and its applications. To a physicist or a geometer, a vector is a directed line segment. To an algebraist or to many engineers, a vector is a list of numbers. To users of more advanced parts of linear algebra, a vector is an element of a vector space. In this latter, most general case, a vector could be any of the above examples, a polynomial, a more general function, or an example of, quite literally, any class of mathematical objects which can be added together and scaled by multiplication.

All too often the meaning in any particular situation is not explained. The result is very confusing to the student. When the word vector appears, a student should make a concerted effort to discover the meaning that is used in the current setting.

When using MATLAB, the situation is clear. A vector is a list of numbers, which may be complex.

```
>> u=[1;2;3;4]
u =
     1
     2
     3
     4
```

To enter a row vector is just as easy.

```
>> v=[1,2,3,4]
v =
     1     2     3     4
```

The *length* of a vector is the number of elements in the list. For example, the length of each of u and v is 4. The MATLAB command `length` will disclose the length of any vector. Try `length(u)` and `length(v)`.

In MATLAB, a vector is just another matrix. Consequently, if your vectors are the same size, then you can add them, as in

```
>> u=[1;2;3;4];v=[5;6;7;8];w=u+v
w =
     6
     8
    10
    12
```

Note that we have suppressed the output with semicolons to save space in the text. However, you might want to try removing the semicolons, as in `u=[1;2;3;4]`,`v=[5;6;7;8]`,`w=u+v`. Try it!

You can also subtract vectors of the same size, as in

```
>> u=[1,2,3];v=[4,5,6];w=u-v
w =
    -3    -3    -3
```

And you can multiply any vector by a scalar.

```
>> v=ones(1,5);w=4*v
w =
     4     4     4     4     4
```

MATLAB's *transpose* operator (a single apostrophe) changes a column vector into a row vector (and vice-versa).

```
>> u=[1;2],v=u'
u =
     1
     2
v =
     1     2
```

Actually, .' is MATLAB's transpose operator and ' is MATLAB's *conjugate transpose* operator. Enter A=[1+i,-2;3i,2-i], then type A' and view the results. Note that rows of A have become columns, but each entry has been replaced with its complex conjugate. Type A.' to appreciate the difference. If each entry of a matrix is a real number, then it doesn't matter whether you use ' or .'.

Linear Combinations of Vectors

Let $\mathbf{v}_1, \mathbf{v}_2, \ldots, \mathbf{v}_p$ be vectors of the same size. Let $x_1, x_2, \ldots, x_p$ be scalars (real numbers)[3]. Then

$$x_1\mathbf{v}_1 + x_2\mathbf{v}_2 + \cdots + x_p\mathbf{v}_p$$

is called a *linear combination* of the vectors $\mathbf{v}_1, \mathbf{v}_2, \ldots, \mathbf{v}_p$.

Example 1. *Consider the vectors*

$$\mathbf{a}_1 = \begin{bmatrix} 2 \\ -2 \\ -2 \end{bmatrix}, \mathbf{a}_2 = \begin{bmatrix} 4 \\ 3 \\ 4 \end{bmatrix}, \text{ and } \mathbf{a}_3 = \begin{bmatrix} -3 \\ 2 \\ 1 \end{bmatrix}.$$

Form and simplify the linear combination $1\mathbf{a}_1 + 0\mathbf{a}_2 - 2\mathbf{a}_3$.

This is easily accomplished in MATLAB.

```
>> a1=[2;-2;-2];a2=[4;3;4];a3=[-3;2;1];
>> 1*a1+0*a2-2*a3
ans =
     8
    -6
    -4
```

The linear combination concept leads to a natural definition for matrix-vector multiplication. One multiplies a matrix A and a vector $\mathbf{x}$ by simply taking a linear combination of the columns of A, using the entries of the vector $\mathbf{x}$ for the scalars in the linear combination.

Example 2. *Let matrix A and vector $\mathbf{x}$ be defined as follows:*

$$A = \begin{bmatrix} 2 & 4 & -3 \\ -2 & 3 & 2 \\ -2 & 4 & 1 \end{bmatrix} \quad \text{and} \quad \mathbf{x} = \begin{bmatrix} 1 \\ 0 \\ -2 \end{bmatrix}.$$

Find the matrix-vector product $A\mathbf{x}$.

---

[3] The scalars could also be complex numbers. Indeed, the vectors themselves could have complex entries.

32

Note that the columns of matrix A are

$$\mathbf{a}_1 = \begin{bmatrix} 2 \\ -2 \\ -2 \end{bmatrix}, \ \mathbf{a}_2 = \begin{bmatrix} 4 \\ 3 \\ 4 \end{bmatrix}, \quad \text{and} \quad \mathbf{a}_3 = \begin{bmatrix} -3 \\ 2 \\ 1 \end{bmatrix},$$

which are identical to the vectors used in Example 1. The matrix-vector product is computed by forming a linear combination of the columns of matrix A, using the entries of vector $\mathbf{x}$ as the scalars. Note that the vector $\mathbf{x}$ has three entries, one for each column of matrix A. Otherwise, the multiplication would not be possible.

$$A\mathbf{x} = \begin{bmatrix} 2 & 4 & -3 \\ -2 & 3 & 2 \\ -2 & 4 & 1 \end{bmatrix} \begin{bmatrix} 1 \\ 0 \\ -2 \end{bmatrix}$$

$$= 1 \begin{bmatrix} 2 \\ -2 \\ -2 \end{bmatrix} + 0 \begin{bmatrix} 4 \\ 3 \\ 4 \end{bmatrix} - 2 \begin{bmatrix} -3 \\ 2 \\ 1 \end{bmatrix}$$

$$= \begin{bmatrix} 8 \\ -6 \\ -4 \end{bmatrix}$$

Note that this is the same solution found in Example 1.

You can perform this same calculation in MATLAB. You can enter the matrix A from scratch, but in this case it is easier to build the matrix A from the vectors just used in Example 1.

```
>> A=[a1,a2,a3]
A =
     2     4    -3
    -2     3     2
    -2     4     1
```

The ability to build matrices in this manner is one of MATLAB's most powerful features. Create the vector $\mathbf{x}$ and perform the multiplication.

```
>> x=[1;0;-2];
>> b=A*x
b =
     8
    -6
    -4
```

Again, it is important to note that this is the same solution found in Example 1. Hence, matrix-vector multiplication is equivalent to taking a linear combination of the columns of the matrix, using the entries of the vector for the scalars.

Example 3. *Solve the following system of equations for x_1, x_2, and x_3.*

$$\begin{aligned} 2x_1 + 4x_2 - 3x_3 &= 8 \\ -2x_1 + 3x_2 + 2x_3 &= -6 \\ -2x_1 + 4x_2 + x_3 &= -4 \end{aligned} \tag{3.1}$$

33

One can certainly argue that the following vectors are equal.

$$\begin{bmatrix} 2x_1 + 4x_2 - 3x_3 \\ -2x_1 + 3x_2 + 2x_3 \\ -2x_1 + 4x_2 + x_3 \end{bmatrix} = \begin{bmatrix} 8 \\ -6 \\ -4 \end{bmatrix} \tag{3.2}$$

But the left side of equation (3.2) can be written as a linear combination.

$$x_1 \begin{bmatrix} 2 \\ -2 \\ -2 \end{bmatrix} + x_2 \begin{bmatrix} 4 \\ 3 \\ 4 \end{bmatrix} + x_3 \begin{bmatrix} -3 \\ 2 \\ 1 \end{bmatrix} = \begin{bmatrix} 8 \\ -6 \\ -4 \end{bmatrix} \tag{3.3}$$

And the left side of equation (3.3) can be written as a matrix-vector product.

$$\begin{bmatrix} 2 & 4 & -3 \\ -2 & 3 & 2 \\ -2 & 4 & 1 \end{bmatrix} \begin{bmatrix} x_1 \\ x_2 \\ x_3 \end{bmatrix} = \begin{bmatrix} 8 \\ -6 \\ -4 \end{bmatrix} \tag{3.4}$$

Equation (3.4) is in the form $A\mathbf{x} = \mathbf{b}$, where

$$A = \begin{bmatrix} 2 & 4 & -3 \\ -2 & 3 & 2 \\ -2 & 4 & 1 \end{bmatrix} \quad \text{and} \quad \mathbf{b} = \begin{bmatrix} 8 \\ -6 \\ -4 \end{bmatrix}.$$

The matrix A is called the *coefficient* matrix. The matrix A is formed by entering the coefficients of x_1, x_2, and x_3 from the system of equations (3.1). One is naively led to explore the possibility of solving the equation $A\mathbf{x} = \mathbf{b}$ by dividing both sides of the equation $A\mathbf{x} = \mathbf{b}$ *on the left* by the matrix A, leading to $A\backslash A\mathbf{x} = A\backslash \mathbf{b}$, or $\mathbf{x} = A\backslash\mathbf{b}$. This is easily accomplished in MATLAB. The matrix A and the vector $\mathbf{b}$ were formed in the previous examples, so all we need to do is enter

```
>> x=A\b
x =
     1
     0
    -2
```

Once again we have demonstrated the ease with which MATLAB can solve mathematical problems. Indeed the method of the previous paragraph works in great generality. **However, it does not always work!** In order for it to work, it is necessary that the coefficient matrix be *nonsingular* (the term nonsingular will be defined in Chapter 9). Unfortunately, many of the applications of linear systems to ordinary differential equations have coefficient matrices which are singular.

Furthermore, since matrix multiplication is a rather complicated procedure, one can expect that matrix division is at least equally complicated. The above naive approach is hiding a lot of mathematics. As a result, we will have to study the problem quite thoroughly. We will return to this in Chapter 9.

Matrix Multiplication

The operation of *matrix multiplication* is little complicated. Fortunately, the matrix-vector form of multiplication demonstrated in Example 2 eases the task.

Matrix Multiplication. *Suppose that matrix A and B have dimensions appropriate for multiplication (the number of columns of A must equal the number of rows of B). Let $\mathbf{b}_1, \mathbf{b}_2, \ldots, \mathbf{b}_p$ represent the columns of matrix B. Then the matrix product AB is defined as follows:*

$$AB = A\left[\mathbf{b}_1, \mathbf{b}_2, \ldots, \mathbf{b}_p\right],$$
$$= \left[A\mathbf{b}_1, A\mathbf{b}_2, \ldots, A\mathbf{b}_p\right].$$

Therefore, the matrix product AB is formed by multiplying each column of the matrix B by the matrix A.

Example 4. *Let A and B represent the following matrices.*

$$A = \begin{bmatrix} 1 & -1 & 1 \\ 0 & 1 & -2 \\ 1 & 0 & 1 \end{bmatrix} \quad \text{and} \quad B = \begin{bmatrix} 2 & 2 & 2 & 2 \\ 2 & 2 & 2 & 2 \\ 2 & 2 & 2 & 2 \end{bmatrix}$$

Find the matrix product AB.

Note that the first column of the matrix B is

$$\mathbf{b}_1 = \begin{bmatrix} 2 \\ 2 \\ 2 \end{bmatrix}.$$

Therefore, the first column of the matrix product AB is $A\mathbf{b}_1$. Note that this multiplication would not be possible if it were not for the fact that the number of rows of B equals the number of columns of A.

$$A\mathbf{b}_1 = \begin{bmatrix} 1 & -1 & 1 \\ 0 & 1 & -2 \\ 1 & 0 & 1 \end{bmatrix} \begin{bmatrix} 2 \\ 2 \\ 2 \end{bmatrix}$$
$$= 2\begin{bmatrix} 1 \\ 0 \\ 1 \end{bmatrix} + 2\begin{bmatrix} -1 \\ 1 \\ 0 \end{bmatrix} + 2\begin{bmatrix} 1 \\ -2 \\ 1 \end{bmatrix}$$
$$= \begin{bmatrix} 2 \\ -2 \\ 4 \end{bmatrix}$$

Since the second, third, and fourth columns of matrix B are identical to the first column of B,

$$A\mathbf{b}_1 = A\mathbf{b}_2 = A\mathbf{b}_3 = A\mathbf{b}_4 = \begin{bmatrix} 2 \\ -2 \\ 4 \end{bmatrix}.$$

Consequently,

$$AB = A\,[\mathbf{b}_1, \mathbf{b}_2, \mathbf{b}_3, \mathbf{b}_4],$$
$$= [A\mathbf{b}_1, A\mathbf{b}_2, A\mathbf{b}_3, A\mathbf{b}_4],$$
$$= \begin{bmatrix} 2 & 2 & 2 & 2 \\ -2 & -2 & -2 & -2 \\ 4 & 4 & 4 & 4 \end{bmatrix}.$$

MATLAB will easily duplicate this effort and relieve us of the drudgery of hand calculations.

```
>> A=[1 -1 1;0 1 -2;1 0 1];B=2*ones(3,4);
>> C=A*B
C =
     2     2     2     2
    -2    -2    -2    -2
     4     4     4     4
```

Remark. *Note that it is possible to define AB only if the number of rows in B equals the number of columns in A. More precisely, If A is an $m \times n$ matrix and B is a $n \times p$ matrix, then the matrix multiplication is possible, and the product AB is a $m \times p$ matrix.*

Mathematicians will say that the matrix multiplication is possible if "the inner dimensions match," and the "outer dimensions give the dimension of the product."

In Example 4, A is 3×3 and B is 3×4. Since the inner dimensions are both 3, it is possible to multiply these matrices. The outer dimensions of matrix A and B predict that the matrix product will have dimensions 3×4, which is precisely what we saw in Example 4.

Example 5. *If*

$$A = \begin{bmatrix} 1 & 2 \\ 3 & 4 \end{bmatrix} \quad and \quad B = \begin{bmatrix} 1 & 0 \\ -1 & 2 \\ 4 & -3 \end{bmatrix},$$

use MATLAB to find the matrix products AB and BA.

Matrix A is 2×2 and matrix B is 3×2. The inner dimensions do not match, so it is not possible to form the matrix product AB. We should expect MATLAB to complain.

```
>> A=[1,2;3,4];B=[1,0;-1,2;4,-3];
>> A*B
??? Error using ==> *
Inner matrix dimensions must agree.
```

However, matrix B is 3×2 and matrix A is 2×2. This time the inner dimensions match, so it is possible to form the product BA. Further, the outer dimensions of B and A predict that the matrix product BA will have dimension 3×2.

```
>> B*A
ans =
     1     2
     5     6
    -5    -4
```

Example 6. *If*

$$A = \begin{bmatrix} 1 & 2 \\ 3 & 4 \end{bmatrix} \quad and \quad B = \begin{bmatrix} 1 & -1 \\ -1 & 1 \end{bmatrix},$$

use MATLAB to find AB and BA and comment on the result.

36

Enter the matrices and perform the calculations.

```
>> A=[1,2;3,4];B=[1,-1;-1,1];
>> A*B,B*A
ans =
    -1     1
    -1     1
ans =
    -2    -2
     2     2
```

This example clearly indicates that matrix multiplication is *not commutative*. In general, $AB \neq BA$. If you switch the order of multiplication, you cannot expect that the answer will be the same. That is why we were careful to divide both sides of the equation $A\mathbf{x} = \mathbf{b}$ on the left to derive the answer $\mathbf{x} = A\backslash \mathbf{b}$ in Example 3.

We will explore further properties of matrices in the exercises.

Array Operations

There is often a need for operations which operate on an element-by-element basis. In MATLAB these operations are built-in and very easy to use. For example, should you need to multiply two vectors on an element-by-element basis, use MATLAB's .* operator.

```
>> v=[1,2,3,4],w=[5,6,7,8],u=v.*w
v =
     1     2     3     4
w =
     5     6     7     8
u =
     5    12    21    32
```

If you look closely, you will see that u is a vector of the same size as v and w, and that each element in u is the product of the corresponding elements in v and w. The MATLAB symbol for this operation is .*, and as you have seen, the result is quite different from what * does. Try entering v*w and see what happens.

The operation we have just defined is called *array multiplication*. There are other array operations. All of them act element-by-element. Try v./w and w./v. This is *array right division*. Then try v.\w and w.\v, and compare the results. This is called *array left division*.

For all array operations it is required that the matrices be of exactly the same size. You might try [1,2;3,4].*[1;1] to see what happens when they are not.

There is one other array operation — *array exponentiation*. As might be expected, this is also an element-by-element operation. The operation A.^2 results in every element of the matrix A being raised to the second power. For example, if A=[1,2;3,4], the command

37

```
>> A=[1,2;3,4],B=A.^2
A =
     1     2
     3     4
B =
     1     4
     9    16
```

raises each entry in A to the second power. Because the command A^2 is equivalent to A*A, you get an entirely different result. Try it.

The built-in MATLAB functions, which we discussed briefly in Chapter 1, are all designed to be *array smart*. This means that if you apply them to a matrix, the result will be the matrix obtained by applying the function to each individual element. For example:

```
>> theta=[0,pi/2,pi,3*pi/2,2*pi],cos(theta)
theta =
         0    1.5708    3.1416    4.7124    6.2832
ans =
    1.0000    0.0000   -1.0000    0.0000    1.0000
```

This is an extremely important feature of MATLAB, as you will discover in the next section.

Plotting in MATLAB

None of these array operations would be important if it were not so easy to create and use vectors and matrices in MATLAB. Here is a typical situation. Suppose we want to define a vector that contains a large number of equally spaced points in an interval $[a, b]$. MATLAB's start:increment:finish construct allows you to generate equally spaced points with ease[4]. For example, the following command generates numbers from 0 to 1 in increments of 0.1:

```
>> t=0:0.1:1
t =
  Columns 1 through 7
         0    0.1000    0.2000    0.3000    0.4000    0.5000    0.6000
  Columns 8 through 11
    0.7000    0.8000    0.9000    1.0000
```

The command y=t.^3 will produce a vector[5] y with 11 entries, each the cube of the corresponding entry in the vector t.

```
>> y=t.^3
y =
  Columns 1 through 7
         0    0.0010    0.0080    0.0270    0.0640    0.1250    0.2160
  Columns 8 through 11
    0.3430    0.5120    0.7290    1.0000
```

---

[4] If you omit the increment, as in t=0:10, MATLAB automatically increments by 1. For example, try q=0:10.

[5] If you prefer column vectors, try t=(0:0.1:1)' and y=t.^3.

We can get a rudimentary plot of the function $y = t^3$ by plotting the entries of y versus the entries of t. MATLAB will do this for us. The command

```
>> plot(t,y)
```

will produce a plot of y versus t in the current figure window. If no figure window exists, then MATLAB will create one for you. MATLAB plots the 11 ordered pairs (t, y) generated by the vectors t and y, connecting consecutive ordered pairs with line segments, to produce an plot similar to the that shown in Figure 3.1.

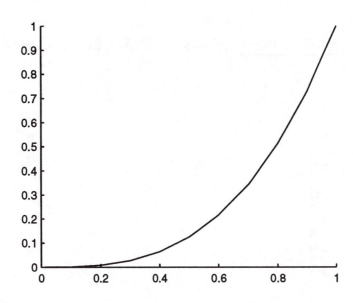

Figure 3.1. A simple plot

You can produce strikingly different plots by choosing different linestyles, markers, and colors. Try the following command and examine its affect[6] on your plot: `plot(t,y,'rx:'),shg`. Experiment with other combinations, such as `plot(t,y,'s')`, `plot(t,y,'md')`, and `plot(t,y,'k--')`. Use the shg command to view the results of each command.

Frequently it is desirable to plot two graphs on the same figure. This is also easily accomplished in MATLAB.

Example 7. *Sketch the graphs of $y = x^2 - 3x + 5$ and $z = x^3 + 6x^2 - 6$ over the interval $[-2, 3]$ on the same figure. Use a solid line type for the first graph and a dashed line type for the second graph.*

---

[6] Type `help plot` at the MATLAB prompt and read the resulting help file. Pay particular attention to the various line types, plot symbols, and colors that can be used with MATLAB's `plot` command. The `shg` command stands for "show the graph," and allows one to preview the results of the current plot command without using the mouse to select the figure window.

You might first try the following.

```
>> x=-2:0.05:3;
>> plot(x,x.^2-3*x+5)
>> plot(x,x.^3+6*x.^2-6,'--'), shg
```

However, the second `plot` command erases the first plot when the second plot is drawn. The following commands will produce an image similar to that in Figure 3.2, with the plots of both graphs on the same axes.

```
>> x=-2:0.05:3;
>> plot(x,x.^2-3*x+5,x,x.^3+6*x.^2-6,'--')
>> grid on, shg
```

The command `grid on` does just what it says — it adds a grid to the figure. Type `grid off` if you want to remove the grid.

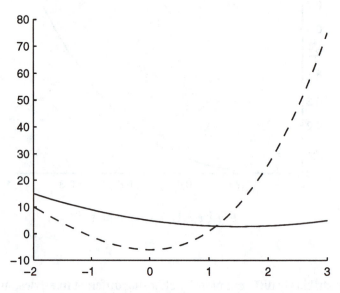

Figure 3.2. Plots of $y = x^2 - 3x + 5$ and $y = x^3 + 6x^2 - 6$.

A second solution involves the commands `hold on` and `hold off`. The command `hold on` tells MATLAB to add subsequent plots to the existing figure, without erasing what is already there. The command `hold off` tells MATLAB to return to the standard procedure of erasing everything before the next plot. This means that `hold on` is in effect until a `hold off` command is executed. Thus, we could have used the following sequence of commands to construct the plot in Figure 3.2.

```
>> x=-2:0.05:3;
>> plot(x,x.^2-3*x+5)
>> hold on
>> plot(x,x.^3+6*x.^2-6,'--')
>> hold off, shg
```

40

Printing Your Plot

After learning how to plot graphs, you will want to print them out. Nothing could be easier in MATLAB. To print what appears in the current figure window[7], simply enter `print` at the MATLAB prompt.

Simple Function M-files

In this section, we will learn how to write simple function M-files. For this purpose, it is necessary to know how to use an editor to create and edit text files on your computer. While the Macintosh and PC versions of MATLAB have a built-in editors (type `edit` at the MATLAB prompt), other versions may not[8]. However, there are a wide variety of editors available, and any of these will do. It is even possible to use a word processor, but if you do it is absolutely essential that you save the file as a text file.

MATLAB has a large variety of built-in mathematical functions. In addition, it is very easy for the user to add to the list. As an example of how this works, let's create a function M-file for the function $f(x) = x^2 - 1$. Open your editor and create a new file containing the following three lines (one of them is blank, and is provided for readability):

```
function y=f(x)

y=x^2-1;
```

The first line of a function M-file must conform to the indicated format. The very first word must be the word `function`. The rest of the first line has the form

```
dependent_variables = function_name(independent_variables)
```

In the function `f`, `y=f(x)` indicates that x is the independent variable and y is the dependent variable.

The rest of the function M-file defines the function using the same syntax as we have been using at the MATLAB prompt. Remember to put a semicolon at the end of lines in which computations are done. Otherwise the results will be printed in the Command Window.

Save the file[9] as `f.m`. That's all there is to it. Now if you want to compute $f(3) = (3)^2 - 1 = 8$, simply enter `f(3)` at the MATLAB prompt.

```
>> f(3)
ans =
    8
```

---

[7] MATLAB's `figure` command allows you to create multiple figure windows. If you have several figure windows open, click any figure window with your mouse to make it the current figure window.

[8] Starting with version 5.2, MATLAB's UNIX version also has a built-in editor.

[9] The file name should always be the same as the function name, followed by the suffix `.m`.

41

Troubleshooting Function M-files. If you don't receive this output, yet you are certain that you have entered the function M-file correctly, it is likely that there is another function M-file called `f.m` that is interfering with the one you just created. If you type `which f` at the MATLAB prompt, MATLAB will respond with the path name of the first directory on the MATLAB path that contains a function M-file called `f.m`. If the path is not the one you expect, you know that you have a name conflict.

You could delete the offending file (probably not a good idea) or you could change the current directory to the directory in which your function M-file `f.m` resides. This can be done from the MATLAB command line. For example, the following command and output

```
>> which f
c:\rewrite\chapter5\f.m
```

indicates that MATLAB will call the function `f.m` from the directory `c:\rewrite\chapter5`. If your version of `f.m` actually resides in the directory `c:\temp`, then the command

```
>> cd c:\temp
```

will ensure that your version of `f.m` will be called first. On a Mac or a PC you can also change directories using the path tool (Type `pathtool` at the MATLAB prompt).

You can check the present working directory with the command

```
>> pwd
ans =
c:\temp
```

You can also check the files in the present working directory with the commands `ls` or `dir`.

```
>> dir

.    ..   f.m
```

There is a very useful MATLAB command which allows a quick check of a function M-file. Enter the command `type f` at the MATLAB prompt. The `type f` command causes MATLAB to print the contents of the function M-file `f.m` in the command window. The output should make it readily apparent what M-file MATLAB is executing. Try it!

Subtle Trouble on Some Platforms. There is one final piece of odd behavior that may occur on some systems. Suppose you make a change to your function (perhaps you change `y = x^2-1;` to `y = 5-x;`). However, when you enter `f(3)`, MATLAB still produces the identical result your previous M-file provided. In addition, the command `type f` does not show any change in the function M-file. In this case MATLAB may be using the function compiled in memory instead of the latest version you just saved on disk. Type `clear f` (or `clear functions` if you want to clear all functions) at the MATLAB prompt to clear the compiled function `f` from memory. If you now enter `f(3)` at the MATLAB prompt, MATLAB is forced to compile the latest function saved to disk.

Making Functions Array Smart. There is one important enhancement you will want to make to the M-file `f.m`. If you try to compute f on a matrix or a vector, you will find that it is not array smart.

```
>> x=1:5
x =
     1     2     3     4     5
>> f(x)
??? Error using ==> ^
Matrix must be square.
```

Edit the function so that it is array smart.

```
function y=f(x)

y=x.^2-1;
```

Now your function can handle matrices.

```
>> x=1:5
x =
     1     2     3     4     5
>> f(x)
ans =
     0     3     8    15    24
```

Notice that if x is a matrix, then so is `x.^2`. On the other hand, 1 is a number, so the difference `x.^2-1` is not defined in ordinary matrix arithmetic. However, MATLAB allows it, and in cases like this will subtract 1 from every element of the matrix. This is a very useful feature in MATLAB.

Naming Functions. Although mathematicians are forever naming their functions f and g, you will probably not want to follow that practice when naming your MATLAB function M-files. For one thing, if you are constantly naming your function M-files `f`, it is inevitable that you will run into frustrating name conflicts. Giving your function M-files distinctive names helps to avoid name conflicts. Secondly, you are going to have some important functions that you will want to save for future use. If you name your function `f`, then overwrite that with a later function also called `f`, the former file is lost, gone forever.

The name of a function M-file is of the form `function_name.m`, where `function_name` is the name you choose to call the function. While this name can be almost anything, there are a few rules.

- It must start with a letter (either upper case or lower case).

- The name must consist entirely of letters, numerals, and underscores (_). No other symbols are allowed. In particular, no periods are allowed.

- The name can be arbitrarily long, but MATLAB will only remember the first 31 characters.

- Do not use names already in use such as `cos`, `plot`, or `dfield5`. If you do, MATLAB will not complain, but you will not be able to use these names for their original purpose until you erase your own files with these names. Should you suspect a conflict of names, use MATLAB's `which` command, as described in **Troubleshooting Function M-files**, to determine which file is being used.

- The variable names used in a function M-file must satisfy the same rules that apply to names of M-files. Other than that they are arbitrary. As a result the file `f.m` could have been written as

```
function stink = funn(skunk)

stink=skunk.^2-1;
```

Save this file as `funn.m` and execute the following command.

```
>> funn(x)
ans =
     0    3    8   15   24
```

Compare this to the output produced earlier with `f(x)` and note that there is no difference. It is important to realize that the variable names in function M-files are *local* to these files; i.e., they are not recognized outside of the M-files themselves. See exercises #10 and #11.

You can use function M-files to plot the graph of a function. You can also call some of MATLAB's powerful numerical routines to analyze the behavior of the function defined by your function M-file.

Example 8. *Find where the graph of $g(x) = x^3 \cos x - 4$ crosses the x-axis on the interval $[-5, 5]$.*

Create a function M-file for the function g and make it array smart.

```
function y=gfunn(x)

y=x.^3.*cos(x)-4;
```

Note that we have decided to name our function `gfunn`. Consequently, you must save this file as `gfunn.m`. Now it is extremely easy to see the graph of `gfunn`. The commands[10]

```
>> x=linspace(-5,5);
>> plot(x,gfunn(x))
>> grid
>> title('y=x^3cos(x)-4')
>> xlabel('x-axis')
>> ylabel('y-axis')
```

will result in a nicely labeled graph of `gfunn` over the interval $[-5, 5]$ (See Figure 3.3).

From Figure 3.3, it can be seen that `gfunn` has a zero near $x = 5$. The precise location of this zero can be located with the MATLAB routine `fzero`. The routine `fzero` requires the name of a function M-file describing the function and the approximate location of the zero. The syntax in this case is `fzero('gfunn',5)`. Notice that the function name must be surrounded by single quotes, and that the `.m` part is not needed.

```
>> z=fzero('gfunn',5)
z =
    4.7497
```

You can check the accuracy of this answer.

---

[10] Type `help linspace` for more information on the `linspace` command.

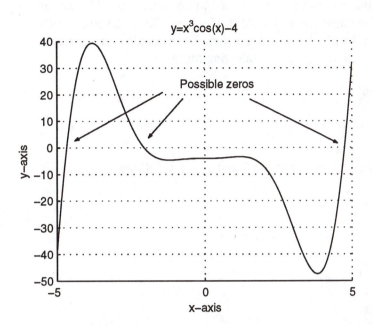

Figure 3.3. Finding a zero of $g(x) = x^3 \cos x - 4$.

```
>> gfunn(z)
ans =
  3.1974e-014
```

Since this is MATLAB's notation for 3.1974×10^{-14}, we see that our answer is very close to a zero of gfunn. You can find the other zeros of the function g by changing the initial guess in the call to fzero. For example, the graph of g in Figure 3.3 indicates that there are zeros near -5 and -2. Try fzero('gfunn',-5) and fzero('gfunn',-2) to find more accurate estimates.

MATLAB allows multiple input and output with its function M-files.

Example 9. *Write a function M-file for the function defined by* $f(t, x) = x^2 - t$.

We need to choose a name for this function, so let's use the mnemonic xsqmt ("x squared minus t"). Then the following M-file describes this function:

```
function y=xsqmt(t,x)

% This is Example 9.
y=x.^2-t;
```

The first line of the file starts with the word function. However, after the word function comes y=xsqmt(t,x), making y the dependent variable, and both t and x independent variables.

Everything on a line in an M-file after a percentage sign (%) is ignored by MATLAB. This can be utilized to put comments in a file, as we have done in the function xsqmt.

45

The comment line can and should be used to document your M-files. It is amazing how quickly we forget why we did things the way we did. Comment lines immediately after the function statement can be read using the MATLAB `help` command. For example, if `xsqmt` is defined as above, saved as `xsqmt.m`, then `help xsqmt` gives the following response.

```
>> help xsqmt

This is Example 9.
```

You can evaluate the function at $t = 2$ and $x = 5$ as follows.

```
>> xsqmt(2,5)
ans =
    23
```

Exercises

1. Consider the vectors

$$\mathbf{v}_1 = \begin{bmatrix} 1 \\ 2 \\ 3 \\ 4 \end{bmatrix}, \quad \mathbf{v}_2 = \begin{bmatrix} -1 \\ 0 \\ 2 \\ -3 \end{bmatrix}, \quad \text{and} \quad \mathbf{v}_3 = \begin{bmatrix} -1 \\ -2 \\ 0 \\ 3 \end{bmatrix}.$$

a) Use MATLAB to compute the linear combination `3*v1-2*v2+4*v3`.

b) Use MATLAB to compute the matrix-vector product `V*c`, where `V=[v1,v2,v3]` and `c=[3;-2;4]`.

c) Compare the results of parts (a) and (b) and explain the importance of this comparison. What does this problem demonstrate?

2. Consider the matrix A and the vectors $\mathbf{b}_1$ and $\mathbf{b}_2$, defined as

$$A = \begin{bmatrix} 3 & 2 & -1 \\ 2 & 0 & -2 \\ -1 & 1 & 3 \end{bmatrix}, \quad \mathbf{b}_1 = \begin{bmatrix} 1 \\ 2 \\ 1 \end{bmatrix}, \quad \text{and} \quad \mathbf{b}_2 = \begin{bmatrix} 3 \\ -1 \\ 4 \end{bmatrix}.$$

a) Use MATLAB to compute `[A*b1,A*b2]`.

b) Use MATLAB to `A*B`, where `B=[b1,b2]`.

c) Compare the results of parts (a) and (b) and explain the importance of this comparison. What does this problem demonstrate?

3. In linear algebra it is shown that the transformation defined by $T(\mathbf{x}) = A\mathbf{x}$ is a *linear transformation*. The linearity depends upon two important facts about matrix-vector multiplication. Consider the matrix A and the vectors $\mathbf{v}_1$ and $\mathbf{v}_2$, defined as

$$A = \begin{bmatrix} -1 & 2 \\ 3 & 5 \end{bmatrix}, \quad \mathbf{v}_1 = \begin{bmatrix} 1 \\ 1 \end{bmatrix}, \quad \text{and} \quad \mathbf{v}_2 = \begin{bmatrix} 3 \\ -1 \end{bmatrix}.$$

Furthermore, let $r = 2$ and $s = 3$.

a) Use MATLAB to show that $A(\mathbf{v}_1 + \mathbf{v}_2)$ is identical to $A\mathbf{v}_1 + A\mathbf{v}_2$.

b) Use MATLAB to show that $A(r\mathbf{v}_1)$ is the same as $r(A\mathbf{v}_1)$ and $(rA)\mathbf{v}_1$.

c) Use MATLAB to show that $A(r\mathbf{v}_1 + s\mathbf{v}_2)$ is the same as $r(A\mathbf{v}_1) + s(A\mathbf{v}_2)$.

4. Consider the matrix

$$A = \begin{bmatrix} 1 & -1 & 2 \\ 2 & 3 & 8 \\ -1 & -2 & 5 \end{bmatrix}.$$

   Form a diagonal matrix with the command D=diag([2,3,4]).

   a) Explain what happens when you multiply matrix A *on the right* by the matrix D.

   b) Explain what happens when you multiply matrix A *on the left* by the matrix D.

5. Use the technique of Example 3 to solve the system

$$x_1 + 2x_2 - 3x_3 = 1,$$
$$2x_1 - x_2 + 4x_3 = -2,$$
$$-x_1 - x_2 + 4x_3 = -5.$$

   Use **format rat** so that MATLAB reports the solution in rational format.

6. When you use the technique of Example 3 to solve the system

$$2x + 3y = 6,$$
$$4x + 6y = 24.$$

   MATLAB fails to find a unique solution. Record the error message given in the command window, then sketch the graph of the system to show why MATLAB was unable to find a unique solution.

7. When you use the technique of Example 3 to solve the system

$$2x + 3y = 6,$$
$$4x + 6y = 12.$$

   MATLAB fails to find a unique solution. Record the error message given in the command window, then sketch the graph of the system to show why MATLAB was unable to find a unique solution.

8. A matrix A is *symmetric* if and only if $A^T = A$, where A^T denotes the transpose of matrix A. A matrix A is *skew symmetric* if and only if $A^T = -A$.

   a) Create a random 5×5 matrix R with the command R=rand(5).

   b) Form the matrix A with the command A=R+R.'. Use MATLAB to compare A.' and A, thus verifying that A is symmetric. What patterns are typically present in a symmetric matrix?

   c) Form the matrix A with the command A=R-R.'. Use MATLAB to compare A.' with -A, thus verifying that A is skew-symmetric. What patterns are typically present in a skew-symmetric matrix? What can always be said about the diagonal elements of a skew-symmetric matrix?

9. A matrix A is *Hermitian* if an only if $A^* = A$, where A^* denotes the *conjugate transpose* of the matrix A.

   a) Create a random 4×4 matrix with the command R=rand(4).

   b) Create a symmetric matrix with the command A=R+R.' and an skew-symmetric matrix with the command B=R-R.'.

   c) Form the matrix H with the command H=A+i*B. Use MATLAB to compare H' and H, thus verifying that H is Hermitian. What patterns are typically present in a Hermitian matrix? What can always be said of the diagonal elements of a Hermitian matrix?

10. Variables defined in functions are *local* to the function. To get a feel for what this means, create the function M-file

```
function y=fcn(x)
A=2;
y=A^x;
```

47

Save the file as fcn.m. In MATLAB's command window, enter the following commands.

```
A=3;
x=5;
y=fcn(x)
A^x
```

Explain the discrepancy between the last two outputs. What is the current value of A in MATLAB's workspace?

11. You can make variables *global*, allowing functions access to the variables in MATLAB's workspace. To get a feel for what this means, create the function M-file

```
function y=gcn(x)
global A
A=2;
y=A^x;
```

Save the file as gcn.m. In MATLAB's command window, enter the following commands.

```
global A
A=3;
x=5;
y=gcn(x)
A^x
```

Why are these last two outputs identical? What is the current value of A in MATLAB's workspace?

12. Consider the matrices

$$A = \begin{bmatrix} 1 & 7 \\ 0 & -3 \end{bmatrix}, \quad B = \begin{bmatrix} -3 & 2 \\ 3 & -2 \end{bmatrix}, \quad \text{and} \quad C = \begin{bmatrix} 3 & 4 & 1 \\ -2 & -1 & 4 \\ 0 & -4 & 3 \end{bmatrix},$$

as well as the vectors

$$\mathbf{v} = \begin{bmatrix} 3 \\ 5 \end{bmatrix}, \quad \mathbf{w} = \begin{bmatrix} 2 \\ -9 \end{bmatrix}, \quad \mathbf{x} = [-4 \ 3], \quad \text{and} \quad \mathbf{y} = [0 \ 6 \ -3].$$

Try the following combinations in MATLAB. Which are defined and which are not? Where a combination is not defined, explain why.

```
a ) A*A   b ) A*B   c ) A*C   d ) A.*A   e ) A+B    f ) A+C    g ) A./C   h ) A.\B
i ) A*x   j ) x*A   k ) v*A   l ) x.*A   m ) A.*x   n ) y*C    o ) C*x    p ) C*y
q ) A*v   r ) A*w   s ) x+v   t ) v+w    u ) x*x    v ) v.*w   w ) w.*v   x ) y.^2
y ) y^2   z ) A^2   aa ) x+y  bb ) v*y   cc ) v.*y  dd ) v*w   ee ) x*w   ff ) x.*w
```

13. The matrix

$$I = \begin{bmatrix} 1 & 0 \\ 0 & 1 \end{bmatrix}$$

is called the 2×2 identity matrix. Replay the command A=rand(2),I=eye(2),A*I,I*A several times to infer that the identity matrix commutes with any 2×2 matrix. *Note: Use the up-arrow to replay a command or cycle through your command history.* Prove your conjecture.

14. Show that the matrices $A = \begin{bmatrix} 1 & -2 \\ 0 & 3 \end{bmatrix}$ and $B = \begin{bmatrix} -2 & -3 \\ 0 & 1 \end{bmatrix}$ commute.

48

15. The MATLAB command `A=round(10*rand(2))` will generate a random 2×2 matrix whose entries are random integers between 0 and 10. Generate two such matrices A and B and compare A*B and B*A. Do this ten times and report how many times A*B = B*A. This can be done easily in MATLAB by using the single composite command

    `A=round(10*rand(2));B=round(10*rand(2));A*B-B*A`

    After executing it once, it can be repeated by using the up arrow to bring it back to the command line. Compare your results with other users.

16. What happens in the previous exercise if you use `.*` instead of `*`?

17. Why is there no `.+` command in MATLAB?

18. On the same figure plot $y = \cos(x)$ and $z = \sin(x)$ over the interval $[0, 4\pi]$. Use different line styles for each curve, and label the figure appropriately.

19. On the same figure plot the three curves $y = \sin(x)$, $y = x - x^3/6$, and $y = x - x^3/6 + x^5/120$ over the interval $[-3, 3]$. Use different line styles for each curve, and label the figure appropriately. Do you recognize the relationship between these three functions?

20. On the same figure plot the graphs of the function $y = e^x$ and its Taylor approximations of order 1, 2, and 3 over the interval $[-3, 3]$. Use different line styles for each curve, and label the figure appropriately.

21. Consider the functions $y_1 = x$, $y_2 = x^2$, and $y_3 = x^4$ on the interval $[0.1, 10]$. Plot these three functions on the same figure using the command `plot`. Now do the same thing with the other plotting commands `semilogx`, `semilogy`, and `loglog`. Turn in only the one that you think is most revealing about the relationship between these functions. Use different line styles for each curve, and label the figure appropriately. (Plotting more than one curve on a figure using any of these commands follows the same procedure used with `plot`.)

22. In three dimensions plot the curve defined by

$$x = t \cos(t),$$
$$y = t \sin(t),$$
$$z = t,$$

over the interval $t \in [0, 4\pi]$ with the `plot3` command. Label the figure appropriately.

23. Write a function M-file to calculate the function $f(x) = \sin(x^2) - 2\cos(x)$. Use it to graph f over the interval $(0, 5)$. Find all of the zeros of f in the interval $(0, 5)$.

24. Type `help quad` at the MATLAB prompt. Read the helpfile, then find a numerical approximation of

$$\int_0^5 \left(\sin(x^2) - 2\cos(x) \right) \, dx.$$

25. Type `help fmins` at the MATLAB prompt. Read the helpfile, then find the minimum value of the function $f(x) = \sin(x^2) - 2\cos(x)$ on the interval $(0, 5)$. At what x-value does this minimum occur?

26. Write a function M-file to calculate the function $g(t, x) = e^{-t}x^2$. Use it to find an approximate numerical value for g at the points $(0, 1)$, $(1, 2)$, $(-1, 1)$, and $(-2, 3)$.

27. Write a function M-file to calculate $f(z) = \sqrt{|z|}$. Use it to graph f over the interval $(-3, 2)$.

28. Write a function M-file to compute the function $f(t) = t^{1/3}$. Use it to graph the function over the interval $[-1, 1]$. (This is not as easy as it looks. Read the section on complex arithmetic in Chapter 1.)

29. Write a function M-file to calculate the function $f(t) = e^{-t}(t^2 + 4e^t - 5)$. Use it to graph f over the interval $(-2.5, 3)$.

30. Find all of the zeros in the interval $(-2.5, 3)$ of the function f defined in the previous exercise.

31. For the function funn defined in this chapter find all solutions to the equation $funn(x) = 5$ in the interval $[-5, 5]$.

49

32. Plot the function $f(t) = 5$ over the interval $[0, 2]$. **Remark:** You will find this exercise more difficult than it looks. It is harder to plot constant functions in MATLAB than most nonconstant functions. There are a variety of tricks that will work, but the methods that are most consistent with the ways used for nonconstant functions (and therefore are most useful in programming applications) use the MATLAB commands `size` and `ones`. Use `help` on these commands.

33. Here is a fancy way of plotting several members of the family of solutions of the ODE $y' + y = \sin t$ on the time interval $[0, 4\pi]$. The ODE is linear, and the general solution $y = -(1/2) \cos t + (1/2) \sin t + Ce^{-t}$ is easily captured with the use of an integrating factor. The following code will plot solutions for $C = -10, -9, \ldots, 10$.

```
>> t=linspace(0,4*pi,500);
>> Y=[];
>> for C=-10:10,Y=[Y;-0.5*cos(t)+0.5*sin(t)+C*exp(-t)];end
>> plot(t,Y)
>> axis tight
>> shg
```

In this case, the `plot` command plots each column of the matrix Y versus the vector t. The command `axis tight` eliminates unused space in the plot ("tightens things up") and the command `shg` ("show the graph") allows you to display the current figure window from the command line.

4. Numerical Methods for ODEs

Numerical methods for solving ordinary differential equations are discussed in many textbooks. Here we will discuss how to go about using some of them in MATLAB. In particular, we will examine how a reduction in the "step size" used by a particular algorithm reduces the error of the numerical solution, but only at a cost of increased computation time. In line with the philosophy that we are not emphasizing programming in this manual, MATLAB routines for these numerical methods will be made available.

Our discussions of numerical methods will be very brief and incomplete. The assumption is that the reader is using a textbook in which these methods are described in more detail.

Euler's Method

We will start with Euler's method. In this very geometric method, we go along the tangent line to the graph of the solution to find the next approximation. That is, to find an approximation of the solution to the initial value problem $y' = f(t, y)$, with $y(a) = y_0$, on the interval $[a, b]$, we set $t_0 = a$, we choose a step size h, and then we inductively define

$$t_{k+1} = t_k + h,$$
$$y_{k+1} = y_k + hf(t_k, y_k),$$

for $k = 0, 1, 2, \ldots, N$, where N is large enough so that $t_N \geq b$. This algorithm is available in the MATLAB command $\texttt{eul}$[1].

Example 1. *Use Euler's method to plot the solution of the initial value problem*

$$y' = y + t, \quad y(0) = 1, \tag{4.1}$$

on the interval $[0, 3]$.

Note that initial value problem (4.1) is in the form $y' = f(t, y)$, where $f(t, y) = y + t$. Before invoking the $\texttt{eul}$ routine, you must first write a function M-file for $f(t, y) = y + t$.

```
function yprime=yplust(t,y)
yprime=y+t;
```

Save the function M-file with the name[2] $\texttt{yplust.m}$.

---

[1] The function $\texttt{eul}$, as well as $\texttt{rk2}$ and $\texttt{rk4}$ described later in this chapter, are function M-files which are not distributed with MATLAB. Type $\texttt{help eul}$ to see if they are installed correctly on your computer. If not, see the Preface for instructions on how to obtain them. The M-files defining these commands are printed in the appendix to this chapter for the illumination of the reader.

[2] Our intent here is to avoid files with duplicate names. That is why we have chosen the distinctive name $\texttt{yplust}$ for our function M-file. We hope that readers do not find this naming convention confusing. If it bothers you that the function M-file name $\texttt{yplust}$ does not match the function $f(t, y)$ in the differential equation $y' = f(t, y)$, simply rewrite the function, changing the first line to $\texttt{function yprime=f(t,y)}$ and saving the file as $\texttt{f.m}$. Thereafter, anywhere we reference the file $\texttt{yplust}$, simply replace this with a reference to $\texttt{f}$.

Before continuing, it is always wise to test that your function M-file is returning the proper output. For example, $f(1, 2) = 2 + 1$, or 3.

```
>> yplust(1,2)
 ans = 3
```

If you don't receive this output, check your function yplust for coding errors. If all looks well, and you still aren't getting the proper output, refer to **Trouble Shooting Function M-files** and **Subtle Trouble on Some Platforms** in the section "Simple Function M-files" of Chapter 3.

Now that your function M-file is operational, it's time to invoke the eul routine. The general syntax is

```
[t,y]=eul('yplust',tspan,y0,stepsize)
```

where 'yplust' is the name of the function M-file in single quotes, tspan is the vector [t0,tfinal] containing the initial and final time conditions, y0 is the y-value of the initial condition, and stepsize is the step size to be used in Euler's method. Actually, the step size is an optional parameter. If you were to enter

```
[t,y]=eul('yplust',tspan,y0);
```

the program would choose a default step size equal to (tfinal-t0)/100}.

The output from the eul routine consists of two column vectors t and y representing the collection of the t_k's and the corresponding y_k's of Euler's method. The following commands should produce a solution of initial value problem (4.1) similar to that shown in Figure 4.1.

```
>> [t,y]=eul('yplust',[0,3],1);
>> plot(t,y)
```

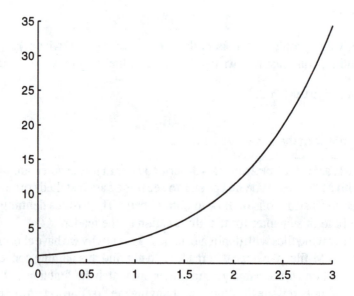

Figure 4.1. Euler's solution of initial value problem (4.1).

Euler's Method Versus the Exact Solution

In this section we want to examine visually the error in the numerical solution obtained from Euler's method.

Example 2. *Re-examine the initial value problem*

$$y' = y + t, \quad y(0) = 1. \tag{4.2}$$

Compare the exact solution and the numerical solution obtained from Euler's method on the interval [0, 3]. *Use a step size of* $h = 0.3$.

The initial value problem in (4.2) is linear and easily solved:

$$y = -t - 1 + 2e^t. \tag{4.3}$$

The numerical solution of the initial value problem in (4.2) is obtained as follows. Note that the step size is $h = 0.3$, as required in the problem statement.

```
>> h=0.3;
>> [teuler,yeuler]=eul('yplust',[0,3],1,h);
```

Use equation (4.3) to compute the exact solution. You will want to choose a finer time increment on the interval [0, 3] so that the exact solution will have the appearance of a smooth curve.

```
>> t=0:.05:3;
>> y=-t-1+2*exp(t);
```

The command

```
>> plot(t,y,teuler,yeuler,'o')
>> legend('Exact','Euler')
```

produces a visual image (Figure 4.2) of the accuracy of Euler's method using the step size $h = 0.3$ (not too good in this case).

Changing the Step Size — Script Files

We would like to repeat the process in Example 2 with different step sizes to analyze graphically how step size affects the accuracy of the approximation.

Example 3. *Examine the numerical solutions provided by Euler's method for the initial value problem*

$$y' = y + t, \quad y(0) = 1,$$

on the interval [0, 3]. *Use step sizes* $h = 0.1, 0.01, 0.001, \ldots$.

53

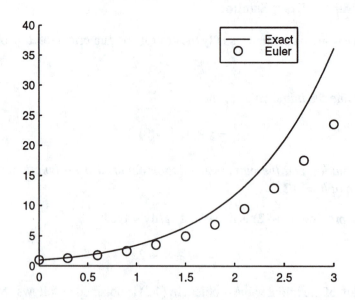

Figure 4.2. Euler versus exact solution (Step size $h = 0.3$).

It will quickly get tedious to type the required commands into the Command Window. Instead let's type them once into an editor and save the file with the name batch1.m, or whatever else you want to call it (as long as it ends in .m). To be precise, this file should contain the following lines:

```
[teuler,yeuler]=eul('yplust',[0,3],1,h);
t=0:.05:3;
y=-t-1+2*exp(t);
plot(t,y,teuler,yeuler,'o')
legend('Exact','Euler')
shg
```

Save the file as batch1.m. This file is an example of a *script file*. A script M-file is a list of MATLAB commands saved as a file with a name ending in .m. Script M-files are very similar to function M-files. In fact, they are simpler. The one obvious difference is that no function statement is needed in a script M-file (as is required in a function M-file). When the name of a script M-file (without the .m) is entered at the MATLAB prompt, the list of commands in the file are executed in the MATLAB workspace in the order of their appearance in the script file. This makes it very easy to execute fairly complicated sequences of commands repeatedly. The following command was used to produce the image in Figure 4.3.

```
>> h=0.1;batch1
```

Repeating this command with various values of h will allow you to examine visually the effect of choosing smaller and smaller step sizes. Try it with step sizes $h = 0.01, 0.001$, etc. Note how the computation time increases as you reduce the step size.

54

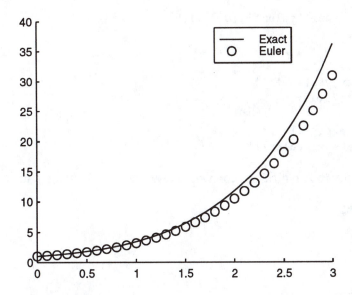

Figure 4.3. Euler versus exact solution (Step size $h = 0.1$).

Further Error Analysis

In this section we further analyze the error inherent in Euler's method. In particular we will examine the *absolute error* at each step, defined as the absolute value of the difference between the actual solution and the numerical solution at that step.

Example 4. *Consider again the initial value problem*

$$y' = y + t, \quad y(0) = 1,$$

on the interval [0, 3]. *Record the maximum error in the Euler's solution for the following step sizes:* $h = 0.2, 0.1, 0.05, 0.025$.

We will record the error at each step by adjusting the script file `batch1.m` from Example 3. We will need to evaluate the exact solution at the same points at which we have the approximate values calculated, i.e. at the points in the vector `teuler`. This is easy. Simply enter `z=-1-teuler+2*exp(teuler)`. Then to compare with the approximate values, we look at the difference of the two vectors `z` and `yeuler`. We are only interested in the magnitude of the error and not the sign. The command `abs(z-yeuler)` gives us a vector with the absolute values of each entry of `z-yeuler`, giving us a vector containing the errors at each step of the Euler computation. Finally, the MATLAB command `maxerror = max(abs(z-yeuler))` finds the largest of all the errors in the Euler computation. We enter this command without the semi-colon, because we want to see the result in the command window. Therefore, to effect these changes we alter the file `batch1.m` as follows:

```
flops(0)
[teuler,yeuler]=eul('yplust',[0,3],1,h);
num_flops=flops
t=0:.05:3;
y=-t-1+2*exp(t);
plot(t,y,teuler,yeuler,'o')
legend('Exact','Euler')
shg
z=-1-teuler+2*exp(teuler);
maxerror=max(abs(z-yeuler))
```

Save this file as `batch2.m`. A typical session would result in the following in the command window:

```
>> h=0.2;batch2
num_flops =
    76
maxerror =
    9.3570
>> h=0.1;batch2
num_flops =
    151
maxerror =
    5.2723
>> h=0.05;batch2
num_flops =
    306
maxerror =
    2.8127
```

Of course, each of these commands will result in a visual display of the error in the Figure Window as well. Thus, we can very easily examine the error in Euler's method both graphically and quantitatively.

If you have executed all of the commands up to now, you will have noticed that decreasing the step size has the effect of reducing the error. But at what expense? The MATLAB command `flops` displays the approximate number of floating point operations required for an operation or sequence of operations. The command `flops(0)` in the first line of `batch2.m` sets the floating point counter to zero. The command `num_flops=flops` provides us with an approximate count of the number of floating point operations required to compute the Euler solution at stepsize h. Note that the number of floating point operations required to compute a solution using Euler's method significantly increases when you lower the step size. Therefore, decreasing the step size used in Euler's method is expensive, in terms of computer time. Notice that halving the step size results in a maximum error almost exactly half the previous one, but requiring almost exactly twice the number of flops.

At this point it would be illuminating to see a plot of how the maximum error changes versus the step size.

Example 5. *Consider again the initial value problem*

$$y' = y + t, \quad y(0) = 1.$$

Sketch a graph of the maximum error in Euler's method versus the step size.

56

To do this using MATLAB, we have to devise a way to capture the various step sizes and the associated errors into vectors, and then use one of the plotting commands. We will call the vector of step sizes h_vect, and the vector of errors err_vect. It is necessary to initialize these vectors. Since at the beginning they contain no information, the proper command is h_vect=[]. The symbol [] stands for the empty matrix; err_vect is initialized in a similar manner.

It will be necessary to add new values of step size to h_vect as they are calculated. The command to do this is h_vect = [h_vect,h] . The result will be a row vector with one more entry than h_vect had previously. We will put all of these commands into a script file.

```
t=0:.05:3;
y=-t-1+2*exp(t);
h_vect=[h_vect,h];
[teuler,yeuler]=eul('yplust',[0,3],1,h);
plot(t,y,teuler,yeuler,'o')
legend('Exact','Euler')
shg
z=-1-teuler+2*exp(teuler);
maxerror=max(abs(z-yeuler));
err_vect=[err_vect,maxerror];
h_vect
err_vect
h=h/2;
```

Save this file as batch3.m. Now we are ready to go. We enter the initialization step and then execute batch3.

```
>> h_vect=[];err_vect=[];h=0.2;
>> batch3
h_vect =
    0.2000
err_vect =
    9.3570
```

When we execute batch3 once more, we get longer vectors.

```
>> batch3
h_vect =
    0.2000    0.1000
err_vect =
    9.3570    5.2723
```

Continuing to execute batch3, we get more and more data. On the seventh iteration we get the following:

```
>> batch3
h_vect =
    0.2000    0.1000    0.0500    0.0250    0.0125    0.0062    0.0031
err_vect =
    9.3570    5.2723    2.8127    1.4548    0.7401    0.3733    0.1875
```

Executing `batch3` one more time, so that we have eight data points in all, we are ready to plot the data. This can be done with any of the plotting commands. Try them all to see which you like best. We will use a loglog graph. Making the graph pretty requires a few more commands:

```
>> loglog(h_vect,err_vect)
>> xlabel('Step size')
>> ylabel('Maximum error')
>> title('Maximum error vs. step size for Euler''s method')
>> grid
```

Figure 4.4 shows the result of these commands.

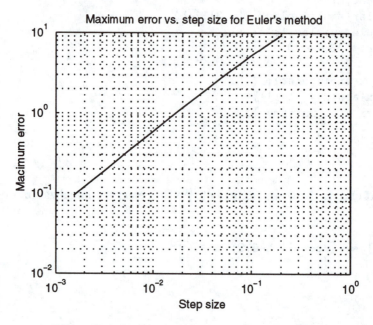

Figure 4.4. Analysis of error in Euler's method.

Figure 4.4 indicates that the graph of the logarithm of the maximum error versus the logarithm of the step size is approximately linear. Indeed, Euler's method is a first order method; i.e., if y_k is the calculated approximation at t_k, then there is a constant C such that

$$|y(t_k) - y_k| \le C|h|, \quad \text{for all } k. \tag{4.4}$$

See Exercises (4–6) for further interpretation of this result.

Hopefully, you now understand two key points about Euler's method: (1) you can increase the accuracy of the solution by decreasing the step size, but (2) you pay for the increase in accuracy with an increase in computing time.

What is needed are solution routines that will provide more accuracy without having to drastically reduce the step size. The Runge-Kutta routines are designed to provide increased accuracy without the drastically increased computing time demanded by Euler's method.

The Second Order Runge-Kutta Method

This method is sometimes called the improved Euler's method. To find this approximation to the solution of the initial value problem $y' = f(t, y)$, with $y(a) = y_0$, on the interval $[a, b]$, we set $t_0 = a$, we choose a step size h, and then we inductively define

$$t_{k+1} = t_k + h,$$
$$s_1 = f(t_k, y_k),$$
$$s_2 = f(t_k + h, y_k + hs_1),$$
$$y_{k+1} = y_k + h(s_1 + s_2)/2,$$

for $k = 0, 1, 2, \ldots, N$ where N is large enough so that $t_N \geq b$. This algorithm is available in the M-file rk2.m, which is listed in the appendix of this chapter.

The syntax for using rk2 is exactly the same as the syntax used for eul. As the name indicates, it is a second order method; i.e., if y_k is the calculated approximation at t_k, then there is a constant C such that

$$|y(t_k) - y_k| \leq C|h|^2, \quad \text{for all } k. \tag{4.5}$$

The error in the second order Runge-Kutta method can be examined experimentally using the same techniques that we illustrated for Euler's method. At this point, we suggest that readers modify the script files, batch1.m, batch2.m, and batch3.m[3], replacing the three letters eul with rk2 everywhere they occur. Of course, you will also want to modify the legends in batch1.m and batch2.m to reflect the fact that you are now using Runge-Kutta 2. Now, replay the commands in Examples 3, 4, and 5 using your newly modified routines.

The Fourth Order Runge-Kutta Method

This is the final method we want to consider. To find this approximation to the solution to the initial value problem $y' = f(t, y)$, with $y(a) = y_0$, on the interval $[a, b]$, we set $t_0 = a$, we choose a step size h, and then we inductively define

$$t_{k+1} = t_k + h,$$
$$s_1 = f(t_k, y_k),$$
$$s_2 = f(t_k + h/2, y_k + hs_1/2),$$
$$s_3 = f(t_k + h/2, y_k + hs_2/2),$$
$$s_4 = f(t_k + h, y_k + hs_3),$$
$$y_{k+1} = y_k + h(s_1 + 2s_2 + 2s_3 + s_4)/6,$$

for $k = 0, 1, 2, \ldots, N$ where N is large enough so that $t_N \geq b$. This algorithm is available in the M-file rk4.m, which is listed in the appendix of this chapter.

---

[3] Renaming the files as batch1rk2.m, batch2rk2.m, and batch3rk2.m is a good idea and some readers will want to make more extensive changes. For example, in batch1rk2.m, you may want your first line to read [trk2,yrk2]=rk2('yplust',[0,3],1,h);, then make a similar change in the plot command.

The syntax for using rk4 is exactly the same as syntax required by eul or rk2. As the name indicates, it is a fourth order method; i.e., if y_k is the calculated approximation at t_k, then there is a constant C such that

$$|y(t_k) - y_k| \le C|h|^4, \quad \text{for all } k. \tag{4.6}$$

The error in the fourth order Runge-Kutta method can be examined experimentally using the same techniques that we illustrated for Euler's and the Runge-Kutta 2 methods. At this point, we suggest that readers modify the script files, batch1.m, batch2.m, and batch3.m (see footnote 3), replacing the three letters eul with rk4 everywhere they occur. Of course, you will also want to modify the legends in batch1.m and batch2.m to reflect the fact that you are now using Runge-Kutta 4. Now, replay the commands in Examples 3, 4, and 5 using your newly modified routines.

Comparing Euler, RK2, and RK4

It is very interesting to compare the accuracy of these three methods, both for individual step sizes, and for a range of step sizes. This can be done by writing script files that are only minor modifications of those already used in this chapter.

Example 6. *Consider again the initial value problem*

$$y' = y + t, \quad y(0) = 1.$$

Sketch a graph of the maximum error in each of the three methods (Euler, RK2, RK4) versus the step size.

Create the following script file.

```
t=0:.05:3;
y=-t-1+2*exp(t);
h_vect=[h_vect,h];
[teuler,yeuler]=eul('yplust',[0,3],1,h);
[trk2,yrk2]=rk2('yplust',[0,3],1,h);
[trk4,yrk4]=rk4('yplust',[0,3],1,h);
plot(t,y,teuler,yeuler,'o',trk2,yrk2,'+',trk4,yrk4,'x')
legend('Exact','Euler','RK2','RK4')
shg
zeuler=-1-teuler+2*exp(teuler);
eulerror=max(abs(zeuler-yeuler));
zrk2=-1-trk2+2*exp(trk2);
rk2error=max(abs(zrk2-yrk2));
zrk4=-1-trk4+2*exp(trk4);
rk4error=max(abs(zrk4-yrk4));
eul_vect=[eul_vect,eulerror];
rk2_vect=[rk2_vect,rk2error];
rk4_vect=[rk4_vect,rk4error];
h_vect
eul_vect
rk2_vect
rk4_vect
h=h/2;
```

Save the file as `batch4.m`. Initialize the following variables at the MATLAB command prompt.

```
>> h=0.1;eul_vect=[];rk2_vect=[];rk4_vect=[];h_vect=[];
```

Execute the script `batch4.m` several[5] times in a row (as you did with `batch3.m` in Example 5) to accumulate data in the four vectors[6]. Then the sequence of commands

```
>> loglog(h_vect,eul_vect,h_vect,rk2_vect,h_vect,rk4_vect)
>> grid
>> xlabel('Step size')
>> ylabel('Maximum error')
>> title('Maximum error vs. step size')
>> gtext('eul')
>> gtext('rk2')
>> gtext('rk4')
```

will result in a plot of the error curves for each of the three methods, as shown in Figure 4.5. Each of the `gtext` commands will put the indicated text at the point on the figure where the mouse button is clicked.

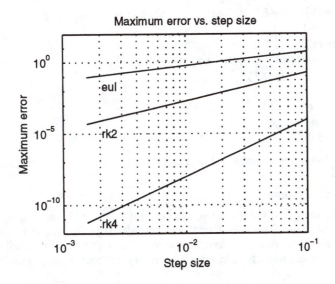

Figure 4.5. Error comparison for Euler, RK2, and RK4.

---

[5] We apologize for the vague use of the word "several" in this directive. Running `batch4` eight consecutive times on a 300MHz machine takes no time at all, but trying to do the same thing on a 50 MHz machine is prohibitively slow. Hence the word "several." Try to run the `batch4` routine as many times as is practical on your platform.

[6] As you execute `batch4.m` repeatedly, you will note that the vector `rk4_vect` appears to have zero entries except for the first two. Is the error from `rk4` actually zero at this point? The answer is no. To see what the entries really are we need to print the vector with more accuracy. To do this, enter `format long` and then `rk4_vect` to see what you get.

Note that Euler's method has the largest error for a given step size, with Runge-Kutta 2 being slightly better, while Runge-Kutta 4 is significantly better than the other two. Also notice that the curves are nearly straight lines in this loglog plot, indicating some sort of power function relationship between the error and and the stepsize, as predicted by the inequalities (4.4), (4.5), and (4.6). The slopes are also roughly comparable to the order of the method in each case (See Exercises (4–6)).

MATLAB's ODE45 Routine

The solver that is recommended to the readers of this book is MATLAB's ode45. This is a sophisticated, variable-step solver used for the numerical solution of initial value problems. Of all the solvers you have looked at thus far, it is the fastest for a prescribed accuracy. It is not necessary to tell ode45 what step size to use, since it uses a variable-step strategy, adjusting the step size at each step of the solution to achieve the needed accuracy and at the same time proceed with the computation as quickly as possible. The routine uses an algorithm developed by J. R. Dormand and P. J. Prince, which is a very clever modification of the Runge-Kutta algorithms.

Readers will be relieved that the calling syntax for ode45 is quite similar to the syntax used by the routines eul, rk2, and rk4. The only difference is ode45 does not require that you enter a stepsize. For example, the following code can be used to compare the exact solution of $y' = y + t$, $y(0) = 1$ with the numerical solution provided by the ode45 command. Try it!

```
>> [tode45,yode45]=ode45('yplust',[0,3],1);
>> t=0:0.05:3;
>> y=-1-t+2*exp(t);
>> plot(t,y,'r',tode45,yode45,'b')
>> legend('Exact','Ode45')
```

Exercises

Each of exercises 1–3 poses an initial value problem on a specific interval. Perform each of the following tasks for each exercise.

i) Find the exact solution.

ii) Examine the error involved in solving this equation numerically using each of the three methods (i.e. eul, rk2, and rk4) with eight different step sizes starting with h_0, and halving each time. On a single sheet of graph paper plot the maximum error vs. step size curve for each of the three methods. A loglog graph would be preferable, but not necessary. Of course you can use MATLAB to do this job for you.

iii) Use MATLAB to plot on a single Graphics Window the graph of the exact solution, together with the plots of the solutions using each of the three methods with step size $h = h_0/4$. Use a distinctive marking for each method. Label the graph appropriately, and print the Graphics Window.

1. $x' = x \sin(3t)$, with $x(0) = 1$ on $[0, 4]$; $h_0 = 0.4$.
2. $y' = (1 + y^2) \cos(t)$, with $y(0) = 0$ on $[0, 6]$; $h_0 = 1$.
3. $z' = z^2 \cos(2t)$, with $z(0) = 1$ on $[0, 6]$; $h_0 = 0.5$.
4. The function with an equation in the form $y = a \cdot x^b$ is called a *power function*. Use MATLAB's plot command to sketch the plot of each of the following power functions on the domain $[0, 2]$. Secondly, use MATLAB's loglog command to plot the function on the domain $[0, 2]$.

   a) $y = 2x^3$
   b) $y = 200x^4$

c) $y = 50x^{-2}$

Describe what the graph of a power function looks like when drawn on a `loglog` graph.

5. If $y = a \cdot x^b$, then $\log_{10} y = b \log_{10} x + \log_{10} a$. For example, the equation $y = 10x^3$ is equivalent to $\log_{10} y = 3 \log_{10} x + \log_{10} 10$. Consequently, if you plot $\log_{10} y$ versus $\log_{10} x$, you should get a line with slope 3 and y-intercept $\log_{10} 10$, or 1. Execute the following code and obtain a printout of the result.

```
x=linspace(0,4);
y=10*x.^3;
plot(log10(x),log10(y))
grid
```

   a) Use a pencil to select two points P and Q on the line in your plot. Estimate the coordinates of the points P and Q and mark these coordinates on your plot with a pencil. Use the slope formula to estimate the slope of the line and record this on your plot.

   b) What is the relationship between the exponent in $y = 10x^3$ and the slope of the line found in part (a)?

6. Execute `h=0.2;eul_vect=[];rk2_vect=[];rk4_vect=[];h_vect=[];` at the MATLAB prompt, then execute `batch4.m` of Example 6 "several" times to regenerate the data set used to plot Figure 4.5. Plot `log(rk4_vect)` versus `log(h_vect)` with the command `plot(log(h_vect),log(rk4_vect), 'ro')`. The points produced by this plot command should appear to maintain a linear relationship. Use the following code to determine and plot the "line of best fit" through the these data points (type `help polyfit` and `help polyval` to get a full explanation of these commands).

```
hold on %hold the prior plot
p=polyfit(log(h_vect),log(rk4_vect),1)
plot(log(h_vect),polyval(p,log(h_vect)),'g')
```

Write the equation of the line of best fit in the form $\log(\text{rk4\_vect}) = C \log(\text{h\_vect}) + B$, where C and B are the slope and intercept captured from the `polyfit` command. Solve this last equation for `rk4_vect`. How closely does this last equation compare with the inequality (4.6)?

   a) Perform a similar analysis on `eul_vect` and `h_vect` and compare with the inequality (4.4).

   b) Perform a similar analysis on `rk2_vect` and `h_vect` and compare with the inequality (4.5).

7. Consider again the initial value problem

$$y' = y + t, \quad y(0) = 1,$$

on the interval [0, 3].

   a) Use trial and error and `batch2.m` with Euler's method to find a step size that yields a maximum error of 1.0000. How many floating point operations are required?

   b) Use trial and error and `batch2.m` with Runge-Kutta 2 to find a step size that yields a maximum error near 1.0000. How many floating point operations are required?

   c) Use trial and error and `batch2.m` with Runge-Kutta 4 to find a step size that yields a maximum error near 1.0000. How many floating point operations are required?

   d) Compare the results of parts (a)–(c). Which appears the most efficient method in terms of floating point operations: `eul`, `rk2`, or `rk4`?

8. The accuracy of any numerical method in solving a differential equation $y' = f(t, y)$ depends on how strongly the equation depends on the variable y. More precisely, the constants that appear in the error bounds depend on the derivatives of f with respect to y. To see this experimentally, consider the two initial value problems

$$y' = y, \quad y(0) = 1,$$
$$y' = e^t, \quad y(0) = 1.$$

You will notice that the two problems have the same solution. For each of the three methods described in this chapter compute approximate solutions to these two initial value problems over the interval [0, 1] using a step size of $h = 0.01$. For each method compare the accuracy of the solution to the two problems.

9. Remember that $y(t) = e^t$ is the solution to the initial value problem $y' = y$, $y(0) = 1$. Then $e = e^1$, and in MATLAB this is e = $exp(1)$. Suppose we try to calculate e approximately by solving the initial value problem, using the methods of this chapter. Use step sizes of the form $1/n$, where n is an integer. For each of Euler's method, the second order Runge-Kutta method, and the fourth order Runge-Kutta method, how large does n have to be to get an approximation e_{app} which satisfies $|e_{app} - e| \leq 10^{-3}$?

10. In the previous problem, show that the approximation to e using Euler's method with step size $1/n$ is $(1+1/n)^n$. As a challenge problem, compute the formulas for the approximations using the two Runge-Kutta methods.

Each of exercises 11–14 poses an intial value problem on a specific interval.

   i) Find the exact solution.

   ii) Use ode45, as shown in the section MATLAB's ODE45 Routine, to find a numerical solution on the indicated interval. Plot the exact and numerical solutions on the same plot. Provide a legend. You may want to experiment with different linestyles to differentiate between the two plots.

11. $y' = \dfrac{y}{t} + t\cos t$, $\quad y(1) = 0$, $\quad [1, 24]$

12. $y' = \dfrac{2t}{2y - 1}$, $\quad y(0) = 1$, $\quad [0, 1]$

13. $y' = y(5 - y)$, $\quad y(0) = 1$, $\quad [0, 1]$

14. $y' = 3 - \dfrac{y}{100 + t}$, $\quad y(0) = 0$, $\quad [0, 100]$

Appendix: M-files for Numerical Methods

In this appendix we provide a listing of the programs eul.m, rk2.m, and rk4.m used in the chapter. The first listing is the file eul.m.

```
function [tout, yout] = eul(FunFcn, tspan, y0, ssize)
%
% EUL    Integrates a system of ordinary differential equations using
%    Euler's method.  See also ODE45 and ODEDEMO.
%    [t,y] = eul('yprime', tspan, y0) integrates the system
%    of ordinary differential equations described by the M-file
%    yprime.m over the interval tspan = [t0,tfinal] and using initial
%    conditions y0.
%    [t, y] = eul(F, tspan, y0, ssize) uses step size ssize
%
% INPUT:
% F      - String containing name of user-supplied problem description.
%          Call: yprime = fun(t,y) where F = 'fun'.
%          t        - Time (scalar).
%          y        - Solution vector.
%          yprime - Returned derivative vector; yprime(i) = dy(i)/dt.
% tspan = [t0, tfinal], where t0 is the initial value of t, and tfinal is
%          the final value of t.
% y0     - Initial value vector.
% ssize - The step size to be used. (Default: ssize = (tfinal - t0)/100).
%
% OUTPUT:
% t    - Returned integration time points (column-vector).
```

```
% y   - Returned solution, one solution row-vector per tout-value.
%
% The result can be displayed by: plot(t,y).

% Initialization

t0=tspan(1);
tfinal=tspan(2);
pm = sign(tfinal - t0);   % Which way are we computing?
if (nargin < 4), ssize = abs(tfinal - t0)/100; end
if ssize < 0, ssize = -ssize; end
h = pm*ssize;
t = t0;
y = y0(:);
tout = t;
yout = y.';

% We need to compute the number of steps.

dt = abs(tfinal - t0);
N = floor(dt/ssize) + 1;
if (N-1)*ssize < dt
  N = N + 1;
end

% Initialize the output.

tout = zeros(N,1);
tout(1) = t;
yout = zeros(N,size(y,1));
yout(1,:) = y.';
k = 1;

% The main loop

while k < N
  if pm*(t + h - tfinal) > 0
    h = tfinal - t;
    tout(k+1) = tfinal;
  else
    tout(k+1) = t0 +k*h;
  end
  k = k+1;
  % Compute the slope
  s1 = feval(FunFcn, t, y); s1 = s1(:); % s1=f(t(k),y(k))
  y = y + h*s1;       % y(k+1) = y(k) + h*f(t(k),y(k))
  t = tout(k);
  yout(k,:) = y.';
end;
```

Since we are not teaching programming, we will not explain everything in this file, but a few things should be explained. The % is MATLAB's symbol for comments. MATLAB ignores everything that appears on a line after a %. The large section of comments that appears at the beginning of the file can be read using the MATLAB help command. Enter help eul and see what happens. Notice that these comments at the beginning take up more than half of the file. That's an indication of how easy it is to program these algorithms. In addition, a couple of lines of the code are needed only to allow the program to handle systems of equations.

Next, we present the M-file for the second order Runge-Kutta method, sans the helpfile, which is pretty much identical to the help file in eul.m. You can type help rk2 to view the help file in MATLAB's command window.

```
function [tout, yout] = rk2(FunFcn, tspan, y0, ssize)
%
% RK2    Integrates a system of ordinary differential equations using
%    the second order Runge-Kutta  method.  See also ODE45 and
%    ODEDEMO.M.
%    [t,y] = rk2('yprime', tspan, y0) integrates the system
%    of ordinary differential equations described by the M-file
%    yprime.m over the interval tspan=[t0,tfinal] and using initial
%    conditions Y0.
%    [t, y] = rk2(F, tspan, y0, ssize) uses step size ssize
%
% INPUT:
% F       - String containing name of user-supplied problem description.
%           Call: yprime = fun(t,y) where F = 'fun'.
%           t       - Time (scalar).
%           y       - Solution column-vector.
%           yprime - Returned derivative column-vector; yprime(i) = dy(i)/dt.
% tspan = [t0, tfinal], where t0 is the initial value of t, and tfinal is
%           the final value of t.
% y0      - Initial value column-vector.
% ssize - The step size to be used. (Default: ssize = (tfinal - t0)/100).
%
% OUTPUT:
% t   - Returned integration time points (column-vector).
% y   - Returned solution, one solution column-vector per tout-value.
%
% The result can be displayed by: plot(t,y).


% Initialization

t0=tspan(1);
tfinal=tspan(2);
pm = sign(tfinal - t0);   % Which way are we computing?
if nargin < 4, ssize = (tfinal - t0)/100; end
if ssize < 0, ssize = -ssize; end
h = pm*ssize;
t = t0;
y = y0(:);

% We need to compute the number of steps.
```

```
dt = abs(tfinal - t0);
N = floor(dt/ssize) + 1;
if (N-1)*ssize < dt
  N = N + 1;
end

% Initialize the output.

tout = zeros(N,1);
tout(1) = t;
yout = zeros(N,size(y,1));
yout(1,:) = y.';
k = 1;

% The main loop

while k < N
  if pm*(t + h - tfinal) > 0
    h = tfinal - t;
    tout(k+1) = tfinal;
  else
    tout(k+1) = t0 +k*h;
  end
  k = k + 1;
  % Compute the slopes
  s1 = feval(FunFcn, t, y); s1 = s1(:);
  s2 = feval(FunFcn, t + h, y + h*s1); s2=s2(:);
  y = y + h*(s1 + s2)/2;
  t = tout(k);
  yout(k,:) = y.';
end;
```

Finally, we have the M-file for the fourth order Runge-Kutta method, without the help file. Type **help rk4** to view the help file in MATLAB's command window.

```
function [tout, yout] = rk4(FunFcn, tspan, y0, ssize)

% RK4   Integrates a system of ordinary differential equations using
%    the fourth order Runge-Kutta  method.  See also ODE45 and
%    ODEDEMO.M.
%    [t,y] = rk4('yprime', tspan, y0) integrates the system
%    of ordinary differential equations described by the M-file
%    yprime.m over the interval tspan=[t0,tfinal] and using initial
%    conditions y0.
%    [t, y] = rk4(F, tspan, y0, ssize) uses step size ssize
%
% INPUT:
% F      - String containing name of user-supplied problem description.
%          Call: yprime = fun(t,y) where F = 'fun'.
%          t      - Time (scalar).
%          y      - Solution column-vector.
```

67

```
%           yprime - Returned derivative column-vector; yprime(i) = dy(i)/dt.
% tspan = [t0, tfinal], where t0 is the initial value of t, and tfinal is
%           the final value of t.
% y0     - Initial value column-vector.
% ssize - The step size to be used. (Default: ssize = (tfinal - t0)/100).
%
% OUTPUT:
% t  - Returned integration time points (column-vector).
% y  - Returned solution, one solution column-vector per tout-value.
%
% The result can be displayed by: plot(t,y).

% Initialization

t0=tspan(1);
tfinal=tspan(2);
pm = sign(tfinal - t0);   % Which way are we computing?
if nargin < 4, ssize = (tfinal - t0)/100; end
if ssize < 0, ssize = -ssize; end
h = pm*ssize;
t = t0;
y = y0(:);

% We need to compute the number of steps.

dt = abs(tfinal - t0);
N = floor(dt/ssize) + 1;
if (N-1)*ssize < dt
  N = N + 1;
end

% Initialize the output.

tout = zeros(N,1);
tout(1) = t;
yout = zeros(N,size(y,1));
yout(1,:) = y.';
k = 1;

% The main loop
while (k < N)
  if pm*(t + h - tfinal) > 0
    h = tfinal - t;
    tout(k+1) = tfinal;
  else
    tout(k+1) = t0 +k*h;
  end
  k = k + 1;
```

```
    % Compute the slopes
    s1 = feval(FunFcn, t, y); s1 = s1(:);
    s2 = feval(FunFcn, t + h/2, y + h*s1/2); s2=s2(:);
    s3 = feval(FunFcn, t + h/2, y + h*s2/2); s3=s3(:);
    s4 = feval(FunFcn, t + h, y + h*s3); s4=s4(:);
    y = y + h*(s1 + 2*s2 + 2*s3 +s4)/6;
    t = tout(k);
    yout(k,:) = y.';
end;
```

5.

Advanced Use of DFIELD5

In this chapter you will explore advanced use of `dfield5`. In particular, you will experiment with differential equations that use step functions and square waves as *forcing functions*. You will also discover that you can write your own function M-files to drive `dfield5`. In later sections you will investigate ordinary differential equations involving a parameter and explore the fundamental changes that occur in the solutions when you subtly perturb the value of the parameter near a critical value (bifurcation point).

Step Functions

The basic unit step function, called the *Heaviside* function, is defined as follows:

$$H(t) = \begin{cases} 0, & \text{if } t < 0; \\ 1, & \text{if } t \geq 0. \end{cases}$$

The Heaviside (unit step) function equals zero for t-values less than zero, one for t-values greater than or equal to zero.

The easiest way to implement the Heaviside function is to use MATLAB's *logical functions*. MATLAB assigns true statements a value of 1 and false statements a value of 0. Confused? Try the following commands at the MATLAB prompt.

```
>> 1<2
ans =
     1
>> 1>=2
ans =
     0
```

In the first case, the inequality $1 < 2$ is a true statement, so MATLAB outputs a 1 in response. In the second case, $1 \geq 2$ is a false statement, so MATLAB outputs a 0 in response. A one means true, a zero means false.

The logical functions in MATLAB are array smart. For example,

```
>> t=-3:3
t =
    -3    -2    -1     0     1     2     3
>> t>=0
ans =
     0     0     0     1     1     1     1
```

The first value of the vector t is -3. Because $-3 \geq 0$ is a false statement, MATLAB places a 0 in the first entry of the response vector. Similarly, because $-2 \geq 0$ and $-1 \geq 0$ are both false statements, MATLAB places zeros in the next two entries of the response vector. However, the fourth value of the vector t is 0 and the $0 \geq 0$ is a true statement, so MATLAB places a 1 in the fourth entry of the response vector. Similarly, $1 \geq 0$, $2 \geq 0$, and $3 \geq 0$ are each true statements, so MATLAB places a 1 in the fifth, sixth, and seventh entries of the response vector.

The logical functions are particularly handy when a problem calls for the use of a *piecewise function*, a function that is defined in two or more pieces.

Example 1. *Sketch the graph of the piecewise defined function*

$$f(t) = \begin{cases} t, & \text{if } t < 1; \\ 1, & \text{if } t \geq 1; \end{cases} \tag{5.1}$$

on the interval [0, 2].

Consider the following pseudocode, where we incorporate the use of MATLAB's logical functions,

$$g(t) = t(t < 1) + 1(t \geq 1).$$

If $t < 1$, then $t < 1$ is a true statement and evaluates to 1; but, $t \geq 1$ is a false statement and evaluates to 0. Consequently, if $t < 1$, then

$$\begin{aligned} g(t) &= t(t < 1) + 1(t \geq 1), \\ &= t(1) + 1(0), \\ &= t. \end{aligned}$$

However, if $t \geq 1$, then $t < 1$ is false and evaluates to 0; but, $t \geq 1$ is now a true statement and evaluates to 1. Consequently, if $t \geq 1$, then

$$\begin{aligned} g(t) &= t(t < 1) + 1(t \geq 1), \\ &= t(0) + 1(1), \\ &= 1. \end{aligned}$$

Note that these results agree with the definition of $f(t)$ in equation (5.1). Consequently, $f(t) = g(t)$ for all values of t and we can use $g(t) = t(t < 1) + 1(t \geq 1)$ to sketch the graph of the piecewise function $f(t)$.

Use logical arrays to create the data for your plot.

```
>> t=0:.05:2;
>> y=t.*(t<1)+1*(t>=1);
```

This last line warrants some explanation. Because t and t<1 are vectors, the array multiplier .* is required in the product t.*(t<1). However, the array multiplier is not required in the scalar-vector product 1*(t>=1) (although you could also use 1.*(t>=1) — try it!). The following command should produce an image similar to that in Figure 5.1.

```
>> plot(t,y,'o-')
```

You might want to experiment with different linestyles and colors. Type help plot to find out how to create other interesting linestyles. For example, try plot(t,y,'r-').

71

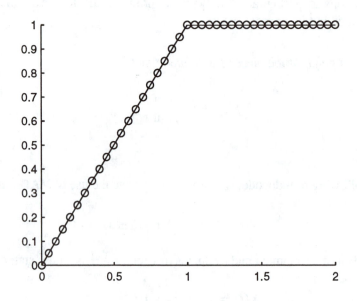

Figure 5.1. The plot of f.

Step Functions in DFIELD5

Now that we know how to implement step functions in MATLAB, let's try to use them in applications requiring the use of dfield5.

Example 2. *An electric circuit containing a resistor and capacitor is called an RC-circuit. Suppose that we have a voltage source*

$$V(t) = \begin{cases} 2, & \text{if } t < 3; \\ 0, & \text{if } t \geq 3; \end{cases} \tag{5.2}$$

driving the circuit. In physics and engineering classes it is shown that the ODE modeling the RC-circuit is given by

$$RC\frac{dv_c}{dt} + v_c = V(t), \tag{5.3}$$

where v_c is the voltage across the capacitor, R is the resistance, and C is the capacitance. Assume that the initial voltage across the capacitor is $v_c(0) = 0$ volts and that the resistance and capacitance are given by $R = 0.2$ ohms and $C = 1$ farad, respectively. Use dfield5 to sketch the solution of this differential equation on the time interval $[0, 6]$.

You can interpret the driving force $V(t)$ in equation (5.2) as a voltage source that is "on" (2 volts) for the first three seconds of the time interval $[0, 6]$, but "off" (0 volts) for the remaining three seconds of this time interval.

The first step is to solve the ODE (3) for dv_c/dt.

$$\frac{dv_c}{dt} = \frac{V(t) - v_c}{RC}$$

72

Pseudocode for the driving force $V(t)$ is $2(t < 3) + 0(t \geq 3)$; or, more simply, $V(t) = 2(t < 3)$. Consequently,

$$\frac{dv_c}{dt} = \frac{2(t < 3) - v_c}{RC} \tag{5.4}$$

is the equation we want to enter in the DFIELD5 Setup window. The right-hand side of equation (5.4) is entered as $(2*(t<3)-v\_c)/(R*C)$, as shown in Figure 5.2.

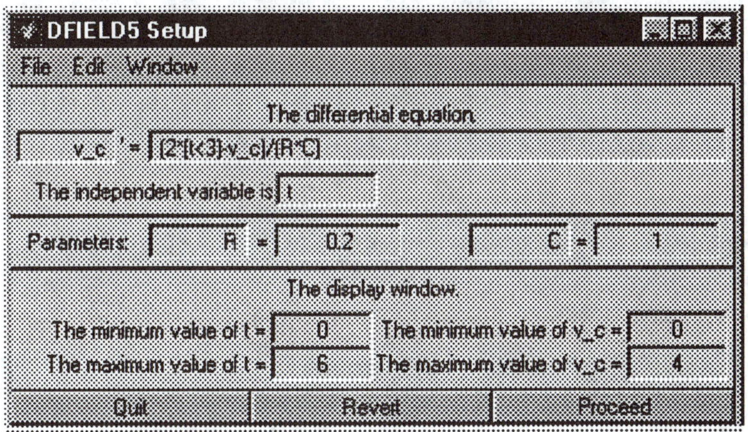

Figure 5.2. Setting up the circuit equation.

Note that the subscript in v_c is entered following an underscore, as in v_c (MATLAB now supports a small subset of TeX commands). Also note that we have introduced and initialized the parameters R and C, as per problem instructions. Set the display window to $0 \leq t \leq 6$, $0 \leq v_c \leq 4$, as shown in the DFIELD5 Setup window in Figure 5.2. Use the DFIELD5 Keyboard input dialog box to compute the trajectory with initial condition $v_c(0) = 0$, as shown in Figure 5.3.

It is interesting to interpret the meaning of the trajectory shown in Figure 5.3. As the power source is turned "on" (2 volts), the voltage across the capacitor increases to a level of 2 volts. At $t = 3$ seconds, the power is turned "off" (0 volts) and the voltage across the capacitor decays exponentially to zero.

Using Function M-files in DFIELD5

You can use your own function M-files in dfield5. The only requirement is that function M-files must be array smart.

Example 3. *Suppose that a fish population is governed by a logistic-like equation*

$$\frac{dP}{dt} = \left(1 - \frac{P}{10}\right) P - H(t), \tag{5.5}$$

where $H(t)$ is a harvesting strategy defined by

$$H(t) = \begin{cases} 3, & \text{if } t < 3; \\ 0.5, & \text{if } t \geq 3; \end{cases} \tag{5.6}$$

73

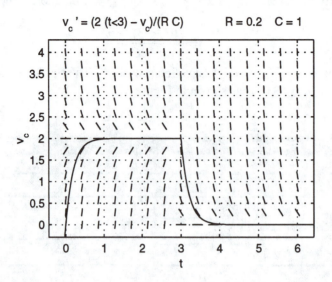

$$v_c' = (2\,(t<3) - v_c)/(R\,C) \qquad R = 0.2 \quad C = 1$$

Figure 5.3. Solution trajectory with initial condition $v_c(0) = 0$.

where P is measured in tons of fish and t is measured in months. Sketch the solution with initial condition $P(0) = 10$. Set the display window so that $0 \le t \le 18$ and $0 \le P \le 25$.

Note that the harvesting strategy in equation (5.6) exhibits a heavy harvesting strategy for the first three months (3 tons per month), while the harvesting is lighter in the remaining months (0.5 tons per month). Pseudocode for $H(t)$ is $3(t < 3) + 0.5(t \ge 3)$. Enter the right hand side of equation (5.5) in the DFIELD5 Setup window as `(1-P/10)*P-(3*(t<3)+0.5*(t>=3))`. Set the display window settings to reflect the problem constraints, $0 \le t \le 18$ and $0 \le P \le 25$. Each of these settings is shown in the image of the DFIELD5 Setup window in Figure 5.4.

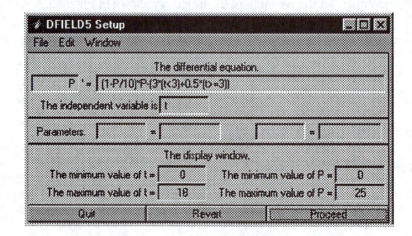

Figure 5.4. Setting up the harvest equation.

Use the DFIELD5 Keyboard input dialog box to compute the solution with initial condition $P(0) = 10$. The resulting solution trajectory is shown in Figure 5.5.

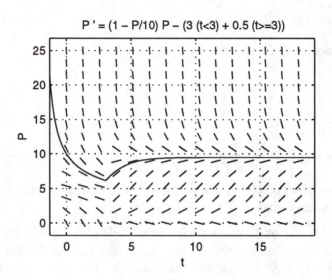

Figure 5.5. Solution with initial condition $P(0) = 10$.

It would appear that the fish population decays exponentially during the first three months of heavy harvesting, then begins to rebound with the lighter harvesting of ensuing months, gradually increasing to a level somewhat less than the initial fish population.

Note that this first approach to the harvest problem is quite similar to the approach taken with the RC circuit in Example 2. Let's look at a second approach that requires the use of a function M-file. Start by opening your editor and creating a function M-file with the following code.

```
function H=harvest(t)
if t<3
    H=3;
else
    H=0.5;
end
```

At first glance this looks like the perfect M-file for our harvesting function $H(t)$. However, this **harvest** function is not array smart, as you can see by entering the following code.

```
>> t=0:6
t =
     0    1    2    3    4    5    6
>> harvest(t)
ans =
    0.5000
```

The **harvest** function does not provide output for each entry in the vector **t**, as it should. The **harvest** function, as written, will give unpredictable results in **dfield5**. The code that drives **dfield5** requires

75

that function M-files used in the DFIELD Setup window must be array smart. Let's rewrite the `harvest` function so that it knows what to do with arrays.

```
function H=harvest(t)
H=3*(t<3)+0.5*(t>=3);
```

Retest the `harvest` function on the vector `t`.

```
>> t=0:6
t =
     0     1     2     3     4     5     6
>> y=harvest(t)
y =
    3.0000    3.0000    3.0000    0.5000    0.5000    0.5000    0.5000
```

Note that this second `harvest` function acted properly on each element of the vector `t`. This rewrite of the `harvest` function is array smart.

Enter the right-hand side of equation (5.5) in the DFIELD5 Setup window of Figure 5.4 as `(1-P/10)*P-harvest(t)`. Note that this is much easier to type than the former `(1-P/10)*P-(3*(t<3)+0.5*(t>=3))` and is sure to eliminate typing mistakes. If you use the DFIELD5 Keyboard input box to compute the trajectory with initial condition $P(0) = 10$, you should produce a plot similar to that shown in Figure 5.5. Try it!

Example 4. *Consider again the harvesting strategy of the fish population in Example 3.*

$$\frac{dP}{dt} = \left(1 - \frac{P}{10}\right) P - H(t) \tag{5.7}$$

However, this time let's propose a different $H(t)$, one that accounts for the fact that harvesting of this population is performed at a constant rate for the first three months of the year, then not at all for the remainder of the year. Suppose that fish are harvested at a constant rate of 3 tons per month for the first three months of the year. During the last nine months of the year, harvesting of the fish population is not allowed. If this same strategy is used for the next three years, discuss the fate of the fish population.

Before using `dfield5` to determine the fate of the fish population, we must first digress and discuss one of the favorite periodic functions of engineers: the *square wave* function. The square wave is a periodic function. During a certain percentage of its period, the square wave function is "on" (equal to 1), and it is "off" (equal to 0) during the remaining portion of its period.

The MATLAB function `sqw`[1] is designed to produce a square wave of arbitrary period and *duty cycle*. The duty cycle is interpreted as a percentage and represents the proportion of time that the square wave is "on." For example, the following code will produce the square wave in a new figure window as shown in Figure 5.6[2].

---

[1] The function M-file `sqw` is not distributed with the normal MATLAB installation. Type `help sqw` at the MATLAB prompt to see if `sqw` is installed on your system. If MATLAB responds with "file not found," read the Preface of this manual for information on how to obtain this file. Or ask your system administrator for help.

[2] The `plot` command always uses the current figure window. Clicking on any figure window with the mouse makes it the current figure window. MATLAB's `figure` command creates a new figure window and makes it the current figure window.

```
>> figure
>> t=linspace(0,12,1000);
>> y=sqw(t,4,25);
>> plot(t,y)
>> set(gca,'xtick',0:12)
```

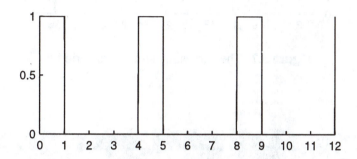

Figure 5.6. Square wave: Period = 4, Duty = 25%

The square wave in Figure 5.6 has period 4 and is "on" (equal to 1) for the first 25% of its period, then "off" (equal to 0) for the remainder of its period. Setting "tick marks" at 0, 1, 2, ..., 12 makes this behavior clearer. Technically, the square wave is a "discontinuous" function, so the vertical lines connecting the "on" and "off" positions should not be present. They are only present because the MATLAB linestyle we selected connects consecutive points of the plot with line segments. Try `plot(t,y,'o')` to get a different perspective. However, the plot in Figure 5.6 provides clear evidence how this function got its name (the square wave).

Our harvesting strategy calls for three tons per month for the first three months of the year, then zero harvesting the remainder of the year. Note that three months is precisely 25% of a year — so the duty cycle is 25%. We want to examine the fish population over three years (36 months). The following code is used to produce the harvesting function pictured in Figure 5.7.

```
>> t=linspace(0,36,1000);
>> y=3*sqw(t,12,25);
>> plot(t,y)
>> set(gca,'xtick',0:3:36)
```

Note that multiplying `sqw` by three is all that is needed to change the amplitude of the square wave. The square wave depicted in Figure 5.7 is exactly the harvesting strategy required by Example 4: fish are harvested at a rate of three tons per month for the first three months of each year. Note that this harvesting strategy repeats itself every 12 months in Figure 5.7.

We have the correct harvesting function. All that is left to do is use this harvesting function in `dfield5` to predict the fate of the fish population. Enter the right hand side of equation (5.7) in the DFIELD5 Setup window as `(1-P/10)*P-3*sqw(t,12,25)`, then set the display window to reflect the passage of 36 months (three years). We used $0 \leq t \leq 36$ and $0 \leq P \leq 25$, as shown in Figure 5.8.

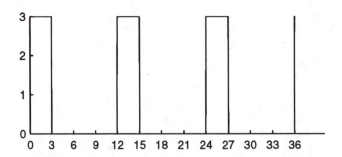

Figure 5.7. The harvesting strategy is periodic

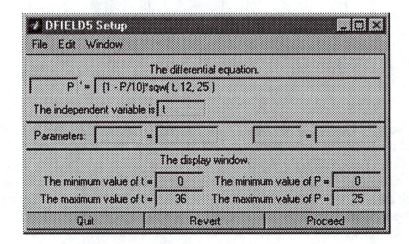

Figure 5.8. Setting up the differential equation.

Finally, use the Keyboard Input window to start a solution trajectory with initial condition $y(0) = 10$ to produce an image similar to that in Figure 5.9.

So, what is the fate of the fish population? When you examine Figure 5.9, the fish population starts at 10 tons, then decays somewhat rapidly during the harvest season (the first three months), then makes a slow recovery to 10 tons over the next nine months. This pattern then repeats itself over the next two years.

Solvers and Solver Settings in DFIELD5

Should you get a plot that appears "kinky" or "jagged," you can adjust the number of plot steps per computation step to smooth your solution trajectory. The Dormand-Prince algorithm implemented in `dfield5` performs interpolation between computation steps. This is accomplished without greatly degrading speed or the efficiency of the algorithm. To change the number of interpolated points between computation steps, select **Options→Solver settings** to open the DFIELD5 Solver settings window, then increase the **Number of plotted points per computation step**. This should smooth most jagged-looking

78

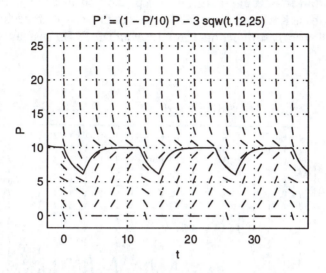

$$P' = (1 - P/10)\, P - 3\,\mathrm{sqw}(t,12,25)$$

Figure 5.9. The fate of the fish population.

curves, unless of course, they are supposed to be jagged.

Dfield5 makes available a number of important solvers. To view the various solvers available in dfield5, select **Options→Solver**. The solvers Euler, Runge-Kutta 2, and Runge-Kutta 4 are identical to the routines eul.m, rk2.m, and rk4.m discussed in Chapter 4. Each of these is a fixed or constant-step algorithm and the user must fix the step size in advance. The Dormand-Prince algorithm is an adaptive algorithm and it adjusts its step size automatically at each step to ensure the greatest possible accuracy. The Dormand-Prince algorithm is selected as the default solver each time your start dfield5.

We can best demonstrate the use of these solvers with an example.

Example 5.

Consider anew the RC-circuit of Example 2. However, in this example, the resistance and capacitance are $R = 2.3$ ohms and $C = 1.2$ farads, respectively. Furthermore, the driving voltage is a square wave with amplitude 3 volts, period 1 second, and duty cycle 25%. Consequently,

$$v_c' = \frac{3\,\mathrm{sqw}(t, 1, 25) - v_c}{RC}.$$

Sketch the solution with initial condition $v_c(0) = 0$.

Enter the equation v_c'=(3*sqw(t,1,25)-v_c)/(R*C), the independent variable t, and the parameters R=2.3 and C=1.2 in the DFIELD5 Setup window. Set the display window so that $0 \le t \le 24$ and $0 \le v_c \le 2$. Click **Proceed** to transfer information to the DFIELD5 Display window.

Select **Options→Solver→Dormand-Prince**. If necessary, reset the Number of plot steps per computation step to 4 and the relative error tolerance to 0.0005 in the DFIELD5 Solver settings window. These are the defaults. Use the Keyboard input window to start the solution trajectory with initial condition $v_c(0) = 0$.

Select **Options→Solver→Runge-Kutta 4**, accept the default settings in the DFIELD5 Solver settings window, then use the Keyboard input window to start the solution trajectory with initial condition $v_c(0) = 0$. If you've followed these directions carefully, your DFIELD5 Display window should resemble that shown in Figure 5.10.

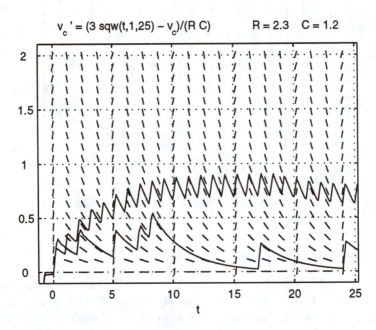

Figure 5.10. Dormand-Prince and Runge-Kutta 4 solutions.

The solution trajectory plotted using the adaptive-step Dormand-Prince algorithm does not look right at all. This can't be correct, particularly in light of the fact that the driving voltage, being a square wave, is periodically supplying a voltage difference every second. Surely, the capacitor should charge, then discharge, charge, then discharge, etc., somewhere on the order of every second.

On the other hand, the solution provided by Runge-Kutta 4, a constant step solver, possesses characteristics one would expect in a solution of this RC-circuit. For example, examine the Runge-Kutta 4 solution over the five second interval [10, 15]. Note that the solution "peaks" five times, once every second, over the interval [10, 15]. After each peak, the solution exhibits decay for almost a full second, then "peaks" again. This is consistent with the fact that the input voltage is a square wave with period one second and equalling 3 volts during the first quarter second of each period. Consequently, the Runge-Kutta 4 solution seems to provide the most realistic solution of the RC circuit.

Why does the less sophisticated solver perform better in this situation? The problem lies with the fact that an adaptive (variable-step) solver, like Dormand-Prince, might take too large a step-size, completely bypassing the point where the input voltage modeled by the square wave suddenly changes from zero to three volts. Select **Options→Solver→Dormand-Prince** then, if necessary, **Options→Solver settings** and note that the current maximum step size is 2.4, which is by default 1/10 of the time interval set in the DFIELD5 Setup window. With the maximum step size set at 2.4 seconds, it is entirely possible that the Dormand-Prince algorithm can skip right over points where the input square wave is "on." A quick look at the graph generated by the Dormand-Prince algorithm in Figure 5.10 suggests that this is exactly what is happening. Because of the large maximum step size, the Dormand-Prince algorithm is "missing" points where the input square wave is "on." Consequently, the graph is not "peaking" every second as it should.

Our square wave input has period one second. However, because the duty cycle is 25%, the square wave is "on" for only 1/4 of a second. If we take 1/10 of the time interval the square wave is "on" each period, and set the maximum step size to 1/40, that should safely ensure that the Dormand-Prince algorithm recognizes all of the points where the square wave input is "on." Set the **maximum step size** in the DFIELD5 Solver settings window equal to 1/40, click the **Change settings** button[3], then use the Keyboard input dialog box to restart a solution trajectory with initial condition $v_c(0) = 0$. Note that this solution trajectory appears to closely mimic the Runge-Kutta 4 solution trajectory.

Actually, it is not a good idea to use a variable step solver in this situation. Variable step solvers can behave quite badly when the input or driving function is discontinuous. There is just too much danger that a variable step solver can "walk" right over a discontinuity without ever seeing that it is there. Even if you choose to use Runge-Kutta 4 in this situation, you should still adjust the step size to the problem constraints. Select **Options→Solver→Runge-Kutta 4**, set the step size to 1/40, then use the Keyboard input dialog box to start a solution trajectory with initial condition $v_c(0) = 0$.

Finally, you might have need to adjust the accuracy of a Dormand-Prince solution by adjusting the **Relative error tolerance** in the DFIELD5 Settings window. Select **Options→Solver→Dormand-Prince** and set the relative error tolerance at 5.0×10^{-8} by entering 5e-8 in the **Relative error tolerance** edit box and clicking the **Change settings** button. If necessary, set the maximum step size back to its default, 2.4 seconds. Use Keyboard input to start a Dormand-Prince solution trajectory at (0, 0) and note the improvement.

One final comment is in order. When you begin adjusting the maximum step size of a variable-step solve like Dormand-Prince, you are essentially changing its character to that of Runge-Kutta or other fixed step solvers. It is probably far wiser to choose a solver appropriate for the job at hand.

Exercises

1. Use your editor to create a Heaviside function.

   ```
   function y=Heaviside(t)
   y=(t>=0);
   ```

   Use this Heaviside function to obtain a plot of the Heaviside function on the interval $[-3, 3]$. Experiment with different linestyles and obtain a printout of your favorite.

2. You can use the Heaviside function defined in exercise 1 to create a number of interesting plots.
   a) Use the command sequence `t=linspace(-3,3,1000);y=Heaviside(t-1);` `plot(t,y,'o')` to create a "delay" of one second in turning "on" the switch represented by the Heaviside function.
   b) Use the command sequence `t=linspace(-3,3,1000);y=Heaviside(-t);plot(t,y,'o')` to reflect the usual graph of the Heaviside function across the vertical axis.

---

[3] Due to a bug in the MATLAB software, the **Change settings** button may not enable. Hit the **Enter** or **Tab** keys, or click the mouse somewhere in the PPLANE5 Solver settings window, and that should enable the **Change Settings** button.

c) Use the command sequence `t=linspace(-3,3,1000);y=Heaviside(t+1)` `-Heaviside(t-1);plot(t,y,'o')` to create a "pulse" of width two and height one, centered at the origin.

d) What sequence of commands will produce the pulse shown in Figure 5.11?

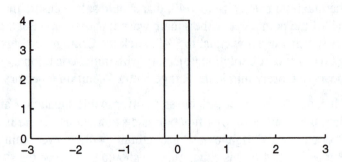

Figure 5.11. A pulse function

3. Use MATLAB to obtain a plot of the *ramp function*

$$R(t) = \begin{cases} 0, & \text{if } t < 0; \\ t, & \text{if } t \geq 0; \end{cases}$$

on the time interval $[-3, 3]$.

4. Write a function M-file that is array smart for

$$f(t) = \begin{cases} -t, & \text{if } t < 0; \\ t^2, & \text{if } t \geq 0. \end{cases}$$

Test your function on the vector `t=-2:2`. Use your function M-file to help obtain a graph of f on the time interval $[-2, 2]$.

5. Obtain a graph of the plot produced by the command sequence

```
t=linspace(0,3);
y=t.*(t<1)+1*(t>=1 & t<2)+t.^2.*(t>=2);
plot(t,y)
```

6. Use MATLAB to obtain a plot of

$$f(t) = \begin{cases} 1, & \text{if } t < 1; \\ 2, & \text{if } 1 \leq t < 2; \\ 3, & \text{if } t \geq 2; \end{cases}$$

on the time interval $[0, 3]$.

7. Write a function M-file that is array smart for

$$h(t) = \begin{cases} 0, & \text{if } t < 0; \\ t, & \text{if } 0 \leq t < 2; \\ 2, & \text{if } 2 \leq t < 4; \\ 6 - t, & \text{if } t \geq 4. \end{cases}$$

82

Test your function on the vector `t=-1:5`. Use your function M-file to help obtain a printout of h on the time interval $[-1, 5]$.

8. An RC-circuit behaves according to

$$RC\frac{dv_c}{dt} + v_c = V(t),$$

where v_c is the voltage across the capacitor. Suppose that the resistance $R = 2.3$ ohms, the capacitance $C = 1.2$ farads, and the driving voltage of the circuit is supplied by

$$V(t) = \begin{cases} 3, & \text{if } t < 5; \\ 0, & \text{if } t \geq 5. \end{cases}$$

Use `dfield5` to sketch the solution of the circuit with initial condition $v_c(0) = 0$.

9. Use the circuit of Exercise 8, with the same resistance and capacitance, but this time the driving voltage is given by

$$V(t) = \begin{cases} 0, & \text{if } t < 5; \\ 3, & \text{if } t \geq 5. \end{cases}$$

Use `dfield5` to sketch the solution of the circuit with initial condition $v_c(0) = 2$. What appears to be the eventual voltage v_c across the capacitor?

10. Create the following function M-file and save it as voltage.m.

```
function v=voltage(t)
v=3*(t>=5);
```

Enter the differential equation of Exercises 8 and 9 in the DFIELD5 Setup window as `v_c'` `=(voltage(t)-v_c)/(R*C)`. Use this equation to duplicate the result in Exercise 9.

11. Assume time is measured in seconds. Use the function M-file `sqw` to obtain plots of a square wave on the time interval $[0, 24]$

   a) with period T=3 and duty cycle 25%.

   b) with period T=2 and duty cycle 50%.

   c) with period T=4 and duty cycle 75%.

   d) Create a plot of a square wave that is "on" for four seconds, "off" for eight seconds, "on" for four seconds, "off" for eight seconds, etc. Obtain a printout of your result on the time period $[0, 96]$.

12. Examine the plot generated by

```
>> t=linspace(0,24,1000);
>> y=sqw(-t,8,50);
>> plot(t,y,'o-')
```

Explain how to use the `sqw` function to obtain a square wave of period four seconds that is "off" for the first second of each period, and "on" for the remaining three seconds in each period. Obtain a printout of your result on the time interval $[0, 24]$.

13. Consider again the RC-circuit of Exercise 8, but this time the voltage source is an amplified square wave, with a period of thirty-two seconds and a duty cycle of 25%. Enter `v_c'=(3*sqw(t,32,25)` `-v_c)/(R*C)` in the DFIELD5 Setup window. Keep the parameters R=2.3 and C=1.2, and adjust the display window so that $0 \leq t \leq 96$ and $0 \leq v_c \leq 4$.

   a) Start a solution trajectory with initial condition $v_c(0) = 0$.

   b) Make the DFIELD5 Display window inactive with **Options→Make the Display Window inactive**. Enter the following code in the MATLAB command window to superimpose the input voltage on the

DFIELD5 Display window.

```
>> t=linspace(0,96,1000);
>> y=3*sqw(t,32,25);
>> plot(t,y)
```

Think of the driving voltage as a signal input into the circuit and the solution trajectory as the resulting output signal. Note that the input signal has been somewhat *attenuated*; that is, the amplitude has decreased in the output signal.

c) Lower the period of the square wave input by entering `v_c'=(3*sqw(t,16,25)-v_c)` `/(R*C)` in the DFIELD5 Setup window. Start a solution trajectory with initial condition $v_c(0) = 0$. Superimpose the square wave as you did in part (b). Note that the output signal is even further attenuated. Repeat this exercise, using periods of 8 and 4 for the input voltage and note how the circuit attenuates input signals with smaller periods (higher frequencies). Why would engineers name this circuit a low pass filter?

14. The plot shown in Figure 5.11

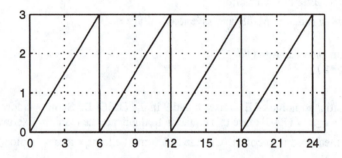

Figure 5.12. A sawtooth wave with period 6, amplitude 3.

is called a *sawtooth wave* and has period six and amplitude three. It was produced with the following MATLAB commands.

```
>> t=linspace(0,24,1000);
>> y=3*rem(t,6)/6;
>> plot(t,y)
>> set(gca,'xtick',0:3:24)
```

Draw sawtooth waves

a) with period 3 and amplitude 1.

b) with period 4 and amplitude 3.

c) with period 8 and amplitude 4.

15. Consider again the *RC*-circuit of Exercise 8, but this time the voltage source is an amplified sawtooth wave, with a period of sixteen seconds. Enter `v_c'=(3*rem(t,32)/32-v_c)/(R*C)` in the DFIELD5 Setup window. Keep the parameters R=2.3 and C=1.2, and adjust the display window so that $0 \le t \le 96$ and $0 \le v_c \le 4$.

a) Start a solution trajectory with initial condition $v_c(0) = 0$.

84

b) Make the DFIELD5 Display window inactive. Enter the following code in the MATLAB command window to superimpose the input voltage on the DFIELD5 Display window.

```
>> t=linspace(0,96,1000);
>> y=3*rem(t,32)/32;
>> plot(t,y)
```

c) Lower the period of the sawtooth wave input by entering `v_c'=(3*rem(t,16,25)-v_c)` `/(R*C)` in the DFIELD5 Setup window. Start a solution trajectory with initial condition $v_c(0) = 0$. Superimpose the sawtooth wave as you did in part (b). Repeat this exercise, using periods of 8 and 4 for the input voltage. Which signals are least attenuated by this circuit?

16. Consider again the fish population from Example 3, where $dP/dt = (1 - P/10)P - H(t)$ described the harvesting of a fish population and the time was measured in months. The harvesting strategy was defined by the piecewise function

$$H(t) = \begin{cases} 3, & \text{if } t < 3; \\ 0.5, & \text{if } t \geq 3. \end{cases}$$

which modeled heavy harvesting in the first three months of each year, lighter harvesting during the remaining months of the year.

The fish population rebounded in Example 3, but this is not always the case. Use the **Keyboard Input** box to help find the critical initial fish population that separates life from extinction.

17. Consider a fish population with constant harvesting, modeled by the equation $dP/dt = (1 - P/10)P - H$. If there is no harvesting allowed, then the fish population behaves according to the logistic model (See Example 4 in Chapter 2).

a) Enter the differential equation in the DFIELD5 Setup window. Add a parameter H and set it equal to zero. Use the **Keyboard input** box to sketch the equilibrium solutions. The equilibrium solutions divide the plane into three regions. Sketch one solution trajectory in each of the three regions.

b) Slowly increase the parameter H and note how the equilibrium solutions move closer together, eventually being replaced by one, then no equilibrium solutions. This is called a *bifurcation* and has great impact on the eventual fate of the fish population. Use `dfield5` to experimentally determine the value of H at which this bifurcation takes place. *Note: If you find it difficult to find the exact bifurcation point with* `dfield5`, *you might consider the graph of* $(1 - P/10)P - H$ *versus P, as in Figure 2.10 in Chapter 2. For what value of H does this plot have exactly one intercept? Check your response in* `dfield5`.

c) Prepare a report on the effect of constant harvesting on this fish population. Your report should include printouts of the DFIELD5 Display window for values of H on each side of the bifurcation point and an explanation of what happens to the fish population for a variety of different initial conditions.

18. One can easily advance or delay a square wave. For example, the MATLAB commands

```
>> t=linspace(0,12,1000);
>> y=sqw(t-1,4,25);
>> plot(t,y)
>> set(gca,'xtick',0:12)
```

produce a square wave with period T=4 and duty cycle d=25%, but the wave is "delayed" one second, as shown in Figure 5.12.

Some might find the term "delay" misleading, as the square wave appears to be moved *forward* one second in time. However, the square pulse does start one second later than usual, hence the term "delay." Use MATLAB to create a plot of a square wave with

a) amplitude three, period $T = 6$ seconds, duty cycle $d = 25\%$, that is delayed 2 seconds on the time interval [0, 24].

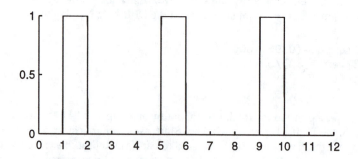

Figure 5.13. A square wave that is delayed one second.

    b) amplitude four, period $T = 12$ months, duty cycle $d = 25\%$, that is delayed 9 months on the time interval $[0, 48]$.

19. Consider again the harvest of the fish population of Example 4, modeled by $dP/dt = (1 - P/10)P - H(t)$, where the time is measured in months. In Example 3, the strategy called for a constant harvesting rate of 3 tons per month, but only during the months of January, February, and March, with no harvesting allowed in the remaining months of the year.

    a) Suppose that the harvesting season is actually during March, April, and May. Use **dfield5** to show the fate of an initial population of ten tons of fish that are harvested at a constant rate of 3 tons per month during harvesting season, but not at all during the remaining months of the year. *Hint: See Exercise 18.*

    b) Show the fate of the population in part (a) if the harvesting season is in November, December, and January of each year.

    c) In part(b), the fate of the fish population depends on the initial population of fish. Use **dfield5** to estimate the critical initial condition that separates survival from extinction.

Appendix: SQW.M

Here is the routine `sqw.m`, used for drawing the square waves in this chapter.

```
function y=sqw(t,T,d)
%SQW Square wave generation
%   sqw(t,T,d) generates a square wave of period T
%   that oscillates between 0 and 1 for the time elements
%   in vector t. The duty cycle determines the percentage
%   of time that the square wave is "on" (equal to 1).
tmp=mod(t,T);
w0=T*d/100;
y=(tmp<w0);
```

6.   Introduction to PPLANE5

A planar system is a system of differential equations of the form

$$x' = f(t, x, y),$$
$$y' = g(t, x, y). \tag{6.1}$$

The variable t in system (6.1) usually represents time and is called the *independent* variable. Frequently, the right-hand sides of the differential equations do not explicitly involve the variable t, and can be written in the form

$$x' = f(x, y),$$
$$y' = g(x, y). \tag{6.2}$$

Such a system is called *autonomous*.

The system

$$x' = y,$$
$$y' = -x, \tag{6.3}$$

is an example of a planar autonomous system. The reader may check, by direct substitution, that the pair of functions $x(t) = \cos t$ and $y(t) = -\sin t$ satisfy both equations and is therefore a solution of system (6.3).

There are a number of ways that we can represent the solution of (6.3) on a graph. The following code was used to construct plots x and y versus t in Figure 6.1.

```
>> t=linspace(0,4*pi);
>> x=cos(t);
>> y=-sin(t);
>> subplot(211)
>> plot(t,x)
>> title('x=cos(t)')
>> xlabel('t')
>> ylabel('x')
>> subplot(212)
>> plot(t,y)
>> title('y=sin(t)')
>> xlabel('t')
>> ylabel('y')
```

You can also plot the solution in the so-called *phase plane*. The commands

```
>> plot(x,y)
>> axis equal
>> title('The plot of y versus x')
>> xlabel('x')
>> ylabel('y')
```

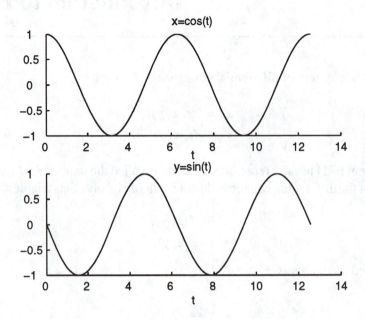

Figure 6.1. Plots of x and y versus t.

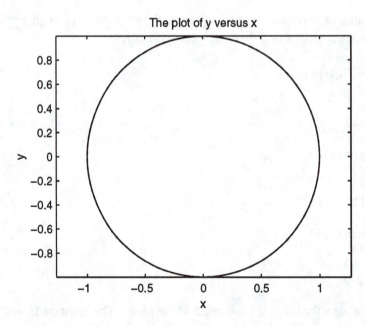

Figure 6.2. The solution in the phase plane.

produce a plot of y versus x in the phase plane, as shown in Figure 6.2.

88

System (6.1) can be written as a vector equation.

$$\begin{bmatrix} x' \\ y' \end{bmatrix} = \begin{bmatrix} f(t, [x, y]^T) \\ g(t, [x, y]^T) \end{bmatrix} \tag{6.4}$$

If we let $\mathbf{x} = [x, y]^T$, then $\mathbf{x}' = [x', y']^T$ and equation (6.4) becomes

$$\mathbf{x}' = \begin{bmatrix} f(t, \mathbf{x}) \\ g(t, \mathbf{x}) \end{bmatrix}. \tag{6.5}$$

Finally, if we define $\mathbf{F}(t, \mathbf{x}) = [f(t, \mathbf{x}), g(t, \mathbf{x})]^T$, then equation (6.5) can be written as

$$\mathbf{x}' = \mathbf{F}(t, \mathbf{x}). \tag{6.6}$$

For a general planar system of the form $\mathbf{x}' = \mathbf{F}(t, \mathbf{x})$, a solution has the vector form $\mathbf{x}(t) = [x(t), y(t)]^T$. The individual components can be plotted as in Figure 6.1, or you can plot in the phase plane as shown in Figure 6.2. The fact that the function $\mathbf{x}(t)$ is a solution of the differential equation $\mathbf{x}' = \mathbf{F}(t, \mathbf{x})$ means that at every point $(t, \mathbf{x}(t))$, the curve $t \rightarrow \mathbf{x}(t)$ must have $\mathbf{F}(t, \mathbf{x}(t))$ as a tangent vector. For a fixed value of t we can imagine the vector $\mathbf{F}(t, \mathbf{x})$ attached to the point $\mathbf{x}$, representing the collection of all possible tangent vectors to solution curves for that specific value of t. Unfortunately this vector field changes as t changes, so this rather difficult visualization is not too useful. However, if the system is autonomous, then system (6.6) can be written

$$\mathbf{x}' = \mathbf{F}(\mathbf{x}) \tag{6.7}$$

and the vector field $\mathbf{F}(\mathbf{x})$ does not change with time t. Therefore, for an autonomous system the same vector field represents all possible tangent vectors to solution curves for all values of t. If a solution curve is plotted parametrically, at each point the vector field must be tangent to the curve.

There is a MATLAB function, pplane5, that makes this visualization easy.[1] This chapter is meant to be an introduction to pplane5, so we will delay discussing some of the more advanced features of pplane5 until you have learned more about systems of differential equations in the ensuing chapters. Actually, the functionality of pplane5 is very similar to that of dfield5, so if you are familiar with dfield5, you will have no trouble with pplane5.

Starting PPLANE5

To see pplane in action, enter pplane5 at the MATLAB prompt. After a short wait, a new window will appear with the label PPLANE5 Setup. Figure 6.3 shows how this window looks on a PC running Windows 95. The appearance might differ slightly depending on your computer, but the functionality will be the same on all machines.

You will notice that there is a rather complicated autonomous system already entered in the upper part of the PPLANE5 Setup window, in the form $x' = f(x, y)$ and $y' = g(x, y)$ of equation (6.2). There is a middle section for entering parameters, although none are entered at the moment. There is another

---

[1] To see if pplane5 is installed properly on your computer enter help pplane5. If it is not installed see the Preface for instructions on how to obtain it.

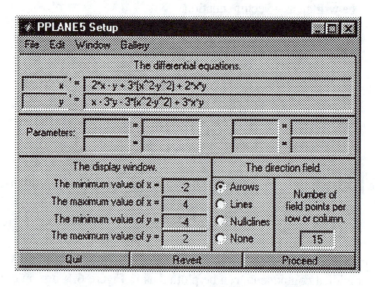

Figure 6.3. The setup window for `pplane5`.

section for describing a "display window," and yet another for defining what kind of field is required. There are are three buttons at the bottom of the window: **Quit, Revert,** and **Proceed.** Across the top of the window are menus: **File, Edit, Window,** and **Gallery.** The **Gallery** menu contains a list of autonomous systems, and to enter one into the PPLANE5 Setup window, it is only necessary to select it from the menu.

We will describe the use of these menus in detail later, but for now leave everything unchanged and click the button labeled **Proceed.** After a few seconds another window will open, this one labeled PPLANE5 Display. An example of this window is shown in Figure 6.4.

The most prominent feature of the PPLANE5 Display window is a rectangle labeled with the variable x on the bottom, and the variable y on the left. The dimensions of this rectangle are slightly larger than the rectangle specified in the PPLANE5 Setup window so as to accomodate the extra space needed by the vectors in the field. Inside this rectangle the PPLANE5 Display window shows the vector field for the equation which is defined in the PPLANE5 Setup window. At each point (x, y) of a grid of points, there is drawn a small arrow. The direction of the vector is the same as that of $\mathbf{F}(x, y) = [f(x, y), g(x, y)]^T$, entered as differential equations in the PPLANE5 Setup window, and the length varies with the magnitude of $\mathbf{F}(x, y)$. This vector must be tangent to any solution curve through (x, y).

There are two buttons on the PPLANE5 Display window, with the labels **Quit** and **Print.** There are menus labeled **File, Edit, Window, Solutions, Graph,** and **Options.** Finally, below the vector field there is a message window through which `pplane` will communicate with the user. At this time it should contain the single word "Ready," indicating that it is ready to follow orders.

To compute and plot a solution curve from an initial point, move the mouse to that point, and click the mouse button. The solution will be computed and plotted, first in the direction in which the independent variable is increasing (the "Forward" direction—see **Options→Solution direction→Forward**), and then in the opposite direction (the "Back" direction—see **Options→Solution direction→Back**). After plotting several solutions, the display will look something like Figure 6.5.

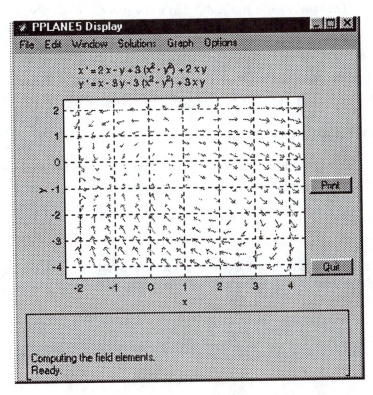

Figure 6.4. The display window for `pplane5`.

$$x' = 2x - y + 3(x^2 - y^2) + 2xy$$
$$y' = x - 3y - 3(x^2 - y^2) + 3xy$$

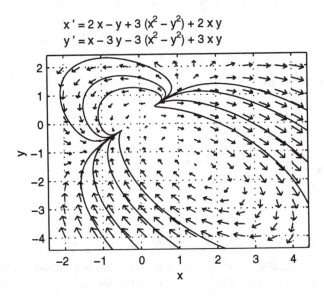

Figure 6.5. Several solutions to the equation.

As in `dfield5`, you can "zoom in" and "zoom back" using the mouse and **Edit**→**Zoom in** and

Edit→Zoom back[3]. The same mouse shortcuts[4] that worked in `dfield5` also work in `pplane5`: on a PC, you can zoom at a point with a right mouse click, and you can hold and drag the right mouse button to draw a "zoom box." There is an addition zoom option, "Zoom square," which we will describe in a later chapter. You can print the PPLANE5 Display window by simply clicking the **Print** button, or you can use one of the options described in Chapter 2, "Printing, Quitting, and Using Clipboards." All of the print and save options that were available for `dfield5` are also available for `pplane5`.

Changing the Differential System — Using the PPLANE5 Setup Window

We will illustrate the use of `pplane` by using it to do a phase plane analysis of the motion of a pendulum. The differential equation for the motion of a pendulum is

$$mL\frac{d^2\theta}{dt^2} = -mg\sin(\theta) - c\frac{d\theta}{dt},\qquad(6.8)$$

where θ is the angular displacement of the pendulum from the vertical, L is the length of the pendulum arm, g is the acceleration due to gravity, and c is the damping constant. If we choose a convenient measure of time by setting $s = \sqrt{g/L}\, t$, the equation becomes[5]

$$\frac{d^2\theta}{ds^2} = -\sin(\theta) - a\frac{d\theta}{ds},\qquad(6.9)$$

where

$$a = \frac{c}{m\sqrt{gL}}$$

is again called the damping constant. This *scaling* of the variables reduces the number of parameters.

We want to write this as a first order system, so we introduce the variables

$$x = \theta \quad \text{and} \quad y = \frac{d\theta}{ds}.\qquad(6.10)$$

Then we have by (6.9) and (6.10)

$$\begin{aligned}
\frac{dx}{ds} &= \frac{d\theta}{ds} = y,\\
\frac{dy}{ds} &= \frac{d^2\theta}{ds^2} = -\sin(\theta) - a\frac{d\theta}{ds} = -\sin(x) - ay,
\end{aligned}\qquad(6.11)$$

or, more simply,

$$\begin{aligned}
x' &= y,\\
y' &= -\sin(x) - ay,
\end{aligned}\qquad(6.12)$$

where the prime indicates differentiation with respect to s. This is a planar autonomous system which we can analyze using `pplane5`. The first thing to do is to enter this system into the PPLANE5 Setup window

---

[3] See Chapter 2, "Zooming and Stopping" for a review.

[4] See the inside cover of this manual for a summary of mouse button actions on various platforms.

[5] Hint: $\frac{d\theta}{dt} = \frac{d\theta}{ds}\frac{ds}{dt} = \sqrt{\frac{g}{L}}\frac{d\theta}{ds}$; thus, $\frac{d^2\theta}{dt^2} = \frac{g}{L}\frac{d^2\theta}{ds^2}$ (why?).

92

by entering the equations into the appropriate boxes, essentially the way they are printed in (6.12). We do need to use an asterisk (∗) to indicate the product of *a* and *y*.

Since the damping constant *a* is not yet a number, we will have to assign it a value in one of the parameter boxes. It does not matter which one. Enter *a* in the left hand side and the value in the right hand side. For the time being, let's use the value $a = 0$. This value can be changed later to a positive number to see the phase plane for a damped pendulum.

Next we have to describe the display rectangle. Since $x = \theta$ represents an angle, and we will want to plot a couple of full periods, we enter $-10 \le x \le 10$. For the dimensions in *y* we have to experiment. For now, $-4 \le y \le 4$ will do.

Finally, we decide which kind of direction field we want and how many field points we want displayed. These are the same options that were available in `dfield`, but are present in the PPLANE5 Setup window. Let's keep the default values.

The completed PPLANE5 Setup window for the pendulum equation is shown in Figure 6.6.

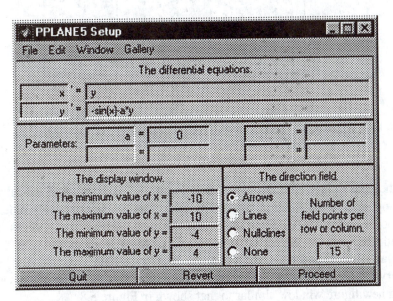

Figure 6.6. The PPLANE5 Setup window for the pendulum equation.

At the very bottom of the PPLANE5 Setup window there are three buttons. Clicking the **Revert** button will return all of the settings in the PPLANE5 Setup window to what they were when you began to make changes. This is useful, for example, when you have made a number of changes and you decide that it would be easiest to start over. The **Quit** button should be used whenever you want to stop using `pplane`. It will close all related windows, and keep MATLAB running.

At this moment, click the **Proceed** button to transfer information from the PPLANE5 Setup window to the PPLANE5 Display window and start the computation of the direction field.

Plotting Solution Curves

Select **Solutions**→**Keyboard input** and start a solution trajectory with initial condition $x(0) = 0$ and $y(0) = 1$. Note that this option behaves in a manner similar to that in `dfield5`. A closed orbit is plotted, giving the phase plane depiction of the standard motion of a pendulum shown in Figure 6.7.

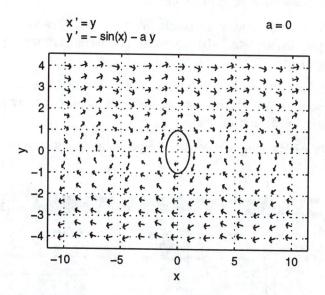

Figure 6.7. Solution with initial condition $(x, y) = (0, 1)$.

Although the phase plane solution pictured in Figure 6.7 is nice, what we might really need are plots of each component of the solution versus time (the phase plane effectively hides time from view). This is easy to do in `pplane5`. Clicking on the **Graph** menu reveals five choices: `x vs. t`, `y vs. t`, `Both`, `3D`, and `Composite`. Select **Graph**→**x vs. t**. Move your mouse over the rectangle in the PPLANE5 Display window and note that it changes to a "crosshairs." Center the "crosshairs" on the solution trajectory pictured in Figure 6.7 and click the mouse button. This action should produce an x versus t plot in a new figure window similar to that shown in Figure 6.8.

Often, we would like to "crop" an x versus t plot, highlighting a particular interval of time. Position the mouse in the PPLANE5 t-plot window, depress the left mouse button, and drag a "cropping box." When you click the **Crop** button, the region in the crop box will expand to the full size of the display axes.

Click the **Go Away** button to close the cropped x versus t plot, then **Go Away** a second time to close the x versus t plot. Select **Graph**→**y vs. t**, then use the mouse to select the solution trajectory in the PPLANE5 Display window shown in Figure 6.7. The result is a plot of the y-component of the solution in Figure 6.7 versus t. At this point, note that it is simpler to change the type of graph by clicking the radio button in the PPLANE5 t-plot window of the type of graph that you wish displayed. The `Both` displays plots of both x and y versus t, while the `3D` button will draw a three dimensional plot, with t being placed on the vertical axis. The `Composite` plot is particularly interesting. First, a three dimensional plot is displayed, then projections of the three dimensional plot are made on three separate planes, the xt-plane, yt-plane, and xy plane.

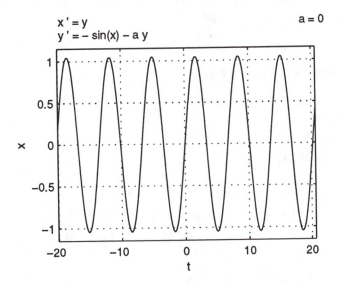

Figure 6.8. An x versus t plot.

Personalizing the Display Window

Just as in `dfield5`, choosing **Edit→Erase all solutions** will delete all solution trajectories from the PPLANE5 Display window. You can place any text you wish in the PPLANE5 Display window by selecting **Edit→Enter text on the Display Window**, typing the desired text in the dialog box, then clicking **OK**. Should you need to delete any graphics object, including text, select **Edit→Delete a graphics object** and click on the object you wish to delete.

You can change the title and axes labels with the commands `title`, `xlabel`, and `ylabel`, just as you did in `dfield5`. For example, make the PPLANE5 Display window the current figure window by clicking on it with your mouse, then enter `xlabel('Angle')` in the MATLAB command window to change the label on the horizontal axis. In a similar manner, try changing the title and the label on the vertical axis.

Saving Systems and Galleries

After spending time entering a planar system, then tweaking the parameters and display window to get things looking just right, you might find yourself wishing for a way to save your system for future use. You might also want to add your system to an existing gallery, or make an entirely new gallery that contains a number of important systems for a lecture or presentation. All of these are possible in `pplane5`.

Consider the second order differential equation

$$my'' + by' + ky = 0, \tag{6.13}$$

modeling an unforced, damped, mass-spring system. The parameters m, b, and k represent the amount of mass, the damping constant, and the spring constant, respectively.

If we want to use `pplane5` to solve this second order equation, we must first write equation (6.13)

95

as a system of two first order differential equations. This is easy to do. First, solve equation (6.13) for y''.

$$y'' = -\frac{b}{m}y' - \frac{k}{m}y \tag{6.14}$$

Next, introduce new variables representing y and y'.

$$\begin{aligned} x_1 &= y \\ x_2 &= y' \end{aligned} \tag{6.15}$$

Then, using equations (6.14) and (6.15),

$$x_1' = y' = x_2,$$
$$x_2' = y'' = -\frac{b}{m}y' - \frac{k}{m}y = -\frac{b}{m}x_2 - \frac{k}{m}x_1,$$

or, more simply,

$$x_1' = x_2,$$
$$x_2' = -\frac{b}{m}x_2 - \frac{k}{m}x_1. \tag{6.16}$$

Select **Edit**→**Clear all** from the **Edit** menu in the PPLANE5 Setup window. Enter system (6.16) in the PPLANE5 Setup window, using x_1 and x_2 for the left hand sides of system (6.16), and x_2 and $-b/m*x\_2-k/m*x\_1$ for the right hand sides, respectively[6]. Set the parameters: m=1, b=2, and c=3. Set the "display window" so that $-5 \le x_1 \le 5$ and $-5 \le x_2 \le 5$. These entries are shown in Figure 6.9.

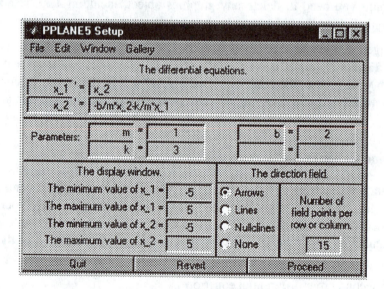

Figure 6.9. The setup window for system (6.16)

---

[6] MATLAB now supports a small subset of TeX commands. The standard way to get a subscript in TeX is to use the code x_1. Superscripts are formed with x^1. If you need more than a single character in a subscript/superscript, use braces, as in x^{2t}.

96

As you can see, it took a lot of effort to set up equation (6.13). Suppose you need to continue your work on this problem at a later date. For example, you may need to present your efforts in tomorrow's class. We need to learn how to save our work.

Select **File→Save the current system** in the PPLANE5 Setup window. You will be given an opportunity to select a folder or directory on your system (perhaps `c:\mysystems`). Once you have done that, enter a filename (perhaps `myspring`) in the Save the system as dialog box. `Pplane5` will automatically attach the suffix `.pps` (pplane system) to your filename. The system is now saved as `myspring.pps` in the folder or directory `c:\mysystem`. When you start your next session of `pplane5`, select **File→Load a system** and select `myspring.pps` from the folder `c:\mysystems`. `Pplane` will load the saved mass-spring system into the PPLANE5 Setup window.

You can also place the current system on the **Gallery** menu by selecting **Gallery→Add current system to the gallery**. You will be asked to provide a name for your system. Name the system `My spring-mass system`, then close System name dialog box by clicking **OK**. If you select **Gallery**, you will note that a separator line has been drawn at the bottom of the gallery menu, below which your system name has been added. Should you delete your system from the PPLANE5 Setup window, then work on some other system for a while, you can return to your system any time be selecting **Gallery**, then clicking on your system name in the **Gallery** menu.

After adding several systems to the **Gallery** menu, you might want to save the gallery for future use (perhaps a lecture or presentation). If you select **File→Save a gallery**, the Gallery selection dialog box will open and allow you to select the systems for inclusion on the saved gallery. Use your platform's technique for selection (on a PC, hold the control key down while selecting the desired systems), then click the **OK** button. The Save gallery as dialog box will open and allow you to select a folder or directory. After selecting a folder or directory, enter a filename for your gallery (`pplane5` will automatically append `.ppg` to your filename) and click **OK** to save the gallery. The next time you start a session of `pplane5`, select **File→Load a gallery**, then load your gallery from the appropriate folder or directory. Note that loading a gallery causes the gallery to be appended below the current gallery. If this is not the behavior you want, select **File→Delete the current gallery** before loading a gallery of your choice.

Take some time to experiment. Add some systems to the current gallery, save the gallery, delete the current gallery, and load some saved galleries. This is one of `pplane5`'s nicest features, and will save you lots of time should you need to present a series of differential equations in a lecture or presentation.

Exercises

1. The planar, autonomous system having the form

$$x' = ax + by,$$
$$y' = cx + dy,$$

where a, b, c, and d are arbitrary real numbers, is called a *linear* system of first order differential equations. Each of the following exercises contains a solution of some linear system. In each exercise, use MATLAB to create a plot of x versus t, y versus t, and a plot of y versus x in the phase plane. Use the `subplot` command, as in

```
subplot(221),plot(t,x),axis tight
subplot(222),plot(t,y),axis tight
subplot(212),plot(x,y),axis equal
```

to produce a plot containing all three plots. Use the suggested time interval.

a) $x = -2e^{-2t} + 3e^{-3t}$
   $y = 4e^{-2t} - 3e^{-3t}$
   $[-0.5, 2]$

b) $x = -4e^{2t} + 3e^{3t}$
   $y = -2e^{2t} + 3e^{3t}$
   $[-2, 0.5]$

c) $x = -2e^{-2t} - e^{3t}$
   $y = e^{-2t} + e^{3t}$
   $[-1.5, 1.0]$

d) $x = \cos 2t - 3 \sin 2t$
   $y = 2 \sin 2t + \cos 2t$
   $[-2\pi, 4\pi]$

e) $x = e^t(\cos t + 5 \sin t)$
   $y = e^t(-8 \sin t + \cos t)$
   $[-\pi, \pi/2]$

f) $x = e^{-t}(\cos t + \sin t)$
   $y = e^{-t}(\cos t - \sin t)$
   $[-\pi, \pi]$

2. Select **Gallery→linear system** in the PPLANE5 Setup window. Adjust the parameters to match the linear system in each of the following exercises. Accept all other default settings in the PPLANE5 Setup window. Select **Solutions→Keyboard input** in the PPLANE5 Display window to start a solution trajectory with the given initial condition. Finally, select **Graph→Both** and click your solution trajectory to obtain plots of x and y versus t. Compare your output to the corresponding problem in Exercise 1.

a) $x' = -4x - y$
   $y' = 2x - y$
   $x(0) = 1, y(0) = 1$

b) $x' = x + 2y$
   $y' = -x + 4y$
   $x(0) = -1, y(0) = 1$

c) $x' = -7x - 10y$
   $y' = 5x + 8y$
   $x(0) = -3, y(0) = 2$

d) $x' = -2x - 4y$
   $y' = 2x + 2y$
   $x(0) = 1, y(0) = 1$

e) $x' = 4x + 2y$
   $y' = -5x - 2y$
   $x(0) = 1, y(0) = 1$

f) $x' = -x + y$
   $y' = -x - y$
   $x(0) = 1, y(0) = 1$

3. In the predator-prey system

$$L' = -L + LP,$$
$$P' = P - LP,$$

L represents a ladybug population and P represents a pest that ladybugs like to eat. Enter the system in the PPLANE5 Setup window, set the display window so that $0 \le L \le 2$ and $0 \le P \le 3$, then select **Arrows** for the direction field.

a) Use the Keyboard input window to start a solution trajectory with initial condition (0.5, 1.0). Note that the ladybug-pest population is periodic. As the ladybug population grows, their food supply dwindles and the ladybug population begins to decay. Of course, this gives the pest population time to flourish, and the resulting increase in the food supply means that the ladybug population begins to grow. Pretty soon, the populations come full cycle to their initial starting position.

b) Suppose that the pest population is harmful to a farmer's crop and he decides to use a poison spray to reduce the pest population. Of course, this will also kill the ladybugs. Question: Is this a wise idea? Adjust the system as follows:

$$L' = -L + LP - HL,$$
$$P' = P - LP - HP.$$

Note that this model assumes that the growth rate of each population is reduced by a fixed percentage of the population. Enter this system in the PPLANE5 Setup window, but keep the original display window settings. Create and set a parameter H=0.25. Start a solution trajectory with initial condition (0.5, 1.0).

c) Repeat part (b) with $H = 0.50, 0.75$, and 1.00. Is this an effective way to control the pests? Why? Describe what happens to each population for each value of the parameter H.

4. Enter the system

$$x' = \sin y - 2 \sin x^2 \sin 2y,$$
$$y' = -\cos x - 2x \cos x^2 \cos 2y,$$

98

in the PPLANE5 Setup window and set the display window so that $-6 \le x \le 6$ and $-10 \le y \le 10$. Select **None** for the vector field.

   a) Select **File→Save the current system** and save the system with the name teddybears.pps.

   b) Select **Gallery→linear system** to load the linear system template. Select **File→Load a system** and load the system teddybears.pps.

   c) Select **Solutions→Keyboard input** and start a solution trajectory with initial condition $(0, \pi/2)$ to create the "legs" of the teddy bears, and a trajectory with initial condition $(-3.5, -7.2)$ to create the "heads" of the teddy bears.

   d) Experiment further with this "wild and wooly" example. Click the mouse in the phase plane to start other solution trajectories. Can you find the "eyes" of the teddy bears?

5. When you portray sufficient solution trajectories in the phase plane so as to determine all of the important behavior of a planar, autonomous system, you have created what is called a "phase portrait." Use pplane5 to create phase portraits for each of the following systems on the indicated display window. *Note: You may want to try ode23s, a "stiff" solver, on some of these systems. Select this solver from the PPLANE5 Display window, using* **Options→Solver→ode23s**.

   a) $x' = y$
      $y' = (1 - y^2)y - x$
      $-3 \le x \le 3, -3 \le y \le 3$

   b) $x' = y + (y^2 - x^2 + 0.5x^4)(1 - x^2)$
      $y' = x(1 - x^2) - y(y^2 - x^2 + 0.5x^4)$
      $-3 \le x \le 3, -2 \le x \le 2$

   c) $x' = x(2y^3 - x^3)$
      $y' = -y(2x^3 - y^3)$
      $-3 \le x \le 3, -3 \le x \le 3$

   d) $x' = y + x^2 + xy$
      $y' = -x + xy + y^2$
      $-2 \le x \le 2, -2 \le x \le 2$

6. The equation $my'' + dy' + ky = 0$ represents a damped, spring-mass system. In the exercises that follow, values of the parameters m, d, and k have been chosen to create a specific example of a damped, spring-mass system with initial position and velocity. Let

$$x_1 = y,$$
$$x_2 = y',$$

and use this change of variables to write each spring-mass system as a planar, autonomous system with initial conditions. Use pplane5 to obtain a printout of the graph of y versus t.

   a) $y'' + 2y' + 2y = 0$
      $y(0) = 0, y'(0) = -1$

   b) $y'' + 4y' + 4y = 0$
      $y(0) = -2, y'(0) = 2$

   c) $y'' + 3y' + 2y = 0$
      $y(0) = 3, y'(0) = 2$

   d) $y'' + 4y = 0$
      $y(0) = -4, y'(0) = -2$

7. Solution flows from tank A into tank B at a rate of 10 gallons per minute, leaving tank B at 10 gallons per minute. Tank A has a 100 gallon capacity and is full. Tank B is also full and has a 200 gallon capacity. Initially, tank A contains 40 pounds of pollutant, tank B contains pure water (no pollutant). Pure water begins pouring into tank A at a rate of 10 gallons per minute. If x_A and x_B represent the number of pounds of pollutant in tank A and tank B, respectively, show that

$$x_A' = -\frac{1}{10}x_A,$$
$$x_B' = \frac{1}{10}x_A - \frac{1}{20}x_B,$$

where $x_A(0) = 40$ and $x_B(0) = 0$.

   a) Use pplane5 to estimate the maximum amount of pollutant in tank B and the time that it occurs.

   b) Use pplane5 to estimate the time that it takes the pollutant in tank B to reach a level of 10 pounds. *Note: This event occurs at two separate times, once as the pollutant level is rising in tank B, and a second time when the pollutant level is decreasing.*

8. Fasten a length of string to a hook. Securely fasten a mass to the the other end of the string. Let θ represent the displacement of the mass from the vertical. Assume that θ is measured in radians, with positive displacements in the counterclockwise direction. Let ω represent the angular velocity of the mass, with positive angular velocity in the counterclockwise direction. Displace the mass 30° counter-clockwise (positive $\pi/6$ radians) from the vertical and release the mass from rest. Let the pendulum decay to a stable equilibrium point at $\theta = 0$, $\omega = 0$.

Important Note. You will not benefit as much as you should from this exercise if you don't first complete parts (a) and (b) before attempting part (c).

a) Without using any technology, sketch graphs of θ versus t and ω versus t.

b) Without using any technology, sketch graphs of ω versus θ. Place ω on the vertical axis, θ on the horizontal axis. *Note: This is a lot harder than it looks. We suggest that you work with a partner or a group and compare solutions before moving on to part (c).*

c) Select **Gallery→pendulum** in the PPLANE5 Setup window. Adjust the damping parameter to D=0.1, and set the display window so that $0 \leq \theta \leq 1$ and $0 \leq \omega \leq 1$. Select **Options→Solution direction→ Forward** and use the Keyboard input window to start a solution trajectory with initial condition $\theta(0) = \pi/6$ and $\omega(0) = 0$. Compare this result with your hand-drawn solution in part (b). Select **Graph→Both** and click your solution trajectory in the phase plane to produce plots of θ versus t and ω versus t. Compare these with your hand-drawn solutions in part (a).

9. Repeat the pendulum experiment of the previous problem. Displace the mass 30° counter-clockwise (positive $\pi/6$ radians) from the vertical, only this time do not release the mass from rest. Instead, push the mass in the clockwise (negative) direction with enough negative angular velocity so that it spins around in a circle exactly one time before settling into a motion decaying to a stable equilibrium. The tricky part of this experiment is the fact that the stable equilibrium point is now $\theta = -2\pi$, $\omega = 0$.

a) Without using any technology, sketch graphs of θ versus t and ω versus t.

b) Without using any technology, sketch graphs of ω versus θ. Place ω on the vertical axis, θ on the horizontal axis. *Note: This is a lot harder than it looks. We suggest that you work with a partner or a group and compare solutions before moving on to part (c).*

c) Select **Gallery→pendulum** in the PPLANE5 Setup window. Adjust the damping parameter to D=0.1, and set the display window so that $-10 \leq \theta \leq 5$ and $-4 \leq \omega \leq 4$. Select **Options→Solution direction→Forward** and use the Keyboard input window to start a solution trajectory with initial condition $\theta(0) = \pi/6$ rad and $\omega(0) = -2.5$ rad/s. Compare this result with your hand-drawn solution in part (b). Select **Graph→Both** and click your solution trajectory in the phase plane to produce plots of θ versus t and ω versus t. Compare these with your hand-drawn solutions in part (a).

10. Consider the predator-prey system

$$R' = R - RF,$$
$$F' = -F + RF,$$

where R and F represent rabbit and fox populations, respectively. Enter the system in `pplane5` and set the display window so that $0 \leq R \leq 2$ and $0 \leq F \leq 2$. Use Keyboard input to start a solution trajectory at $R = 0.5$ and $T = 1$. Note that the trajectory is nearly periodic (actually periodic, but `pplane5` has some estimation difficulties here). Use `pplane5` to estimate the time it takes to travel this periodic trajectory exactly once; i.e., find the period of the oscillation. *Hint: Try cropping an x versus t plot.*

11. Start a new session by exiting and restarting `pplane5`. Select **File→Delete the current gallery** in the PPLANE5 Setup window. Examine your textbook and select three planar, autonomous systems of interest. Enter the first system and set the parameters (if any), then adjust the display window and select the direction field. Select **Gallery→Add current system to the gallery** to add your system to the gallery. Add each of your remaining systems to the gallery, then select **File→Save a gallery** to create and save your new gallery.

a) Start a new session by exiting and restarting `pplane5`. Delete the current gallery, then load your newly created gallery.

b) Check that each system in the gallery works as it should.

7.   Solving ODEs in MATLAB

This is one of the most important chapters in this manual. Here we will describe how to use MATLAB's built-in numerical solvers to find approximate solutions to almost any system of differential equations. At this point readers might be thinking "I've got `dfield5` and `pplane5` and I'm all set. I don't need to know anything further about numerical solvers." Unfortunately, this would be far from the truth. For example, how would we handle a system modeling a driven, damped oscillator such as

$$x_1' = x_2,$$
$$x_2' = -2x_1 - 3x_2 + \cos t. \tag{7.1}$$

System (7.1) is **not autonomous**! Pplane5 can only handle autonomous systems. Furthermore, how would we handle a system such as

$$x_1' = x_1 + 2x_2,$$
$$x_2' = -x_3 - x_4,$$
$$x_3' = x_1 + x_3 + x_4, \tag{7.2}$$
$$x_4' = -2x_2 - x_4.$$

The right-hand sides of system (7.2) do not explicitly involve the independent variable t (the system is autonomous), but system (7.2) is **not planar**! Pplane5 can only handle autonomous systems of two first order differential equations (planar systems).

Finally, in addition to handling systems of differential equations that are neither planar nor autonomous, we have to develop some sense of trust in the output of numerical solvers. How accurate are they?

MATLAB's ODE Suite

With the appearance of version 5 of MATLAB, there appeared a new suite of ODE solvers. There are now seven solvers in the suite: `ode23`, `ode45`, `ode113`, `ode15s`, `ode23s`, `ode23t`, and `ode23tb`. We will spend most of our time talking about `ode45`. The `ode45` solver uses a variable step Runge-Kutta procedure. Six derivative evaluations are used to calculate an approximation of order five, and then another of order four. These are compared to come up with an estimate of the error being made at the current step. It is required that this estimated error at any step should be less than a predetermined amount. This predetermined amount can be changed using two tolerance parameters, and a very high degree of accuracy can be required of the solver. We will discuss this later in this chapter. For most of the uses in this Manual the default values of the tolerance parameters will provide sufficient accuracy, and we will first discuss the use of the solver with the default accuracy.

Ode45 is an excellent general purpose solver that can be used in most cases. We will spend a little time toward the end talking about `ode15s` which is an excellent solver for those systems which are "stiff." Although we will almost exclusively use `ode45`, it is important to realize that the calling syntax is the same for each solver in MATLAB's suite of ODE solvers.

In addition to being very powerful, MATLAB's suite of numerical solvers are easy to use, as the reader will soon discover. The methods described herein are used regularly by engineers and scientists, and are available in any version of MATLAB. The student who learns the techniques described here will find them useful in many later circumstances.

Single First Order Differential Equations

Readers will probably note that we first used `ode45` when discussing numerical routines (See Chapter 5, MATLAB's ODE45 Routine). However, for completeness, we will do another example here. Assuming that our initial value problem has the form $x' = f(t, x)$, $x(t_0) = x_0$, the calling syntax for `ode45` is `ode45('Fcn',tspan,x0)`, where Fcn is the name of your ODE file in single quotes, `tspan=[t0,tfinal]` is a vector containing the initial and final times, and `x0` is the x-value of the initial condition.

Example 1. *Use* `ode45` *to plot the solution of the initial value problem*

$$x' = \frac{\cos t}{2x - 2}, \quad x(0) = 3, \tag{7.3}$$

on the interval $[0, 2\pi]$.

Note that equation (7.3) is in the form $x' = f(t, x)$, where $f(t, x) = \cos t/(2x - 2)$. Open your editor and create the following function ODE file.

```
function xprime=test(t,x)
xprime=cos(t)/(2*x-2);
```

Save the file[1] as `test.m`. Note that the function M-file mimics the equation $x' = \cos t/(2x - 2)$. Before going further, it is always a good idea to test your function M-file. Note that $f(\pi, 2) = \cos(\pi)/(2(2) - 2) = -1/2$.

```
>> test(pi,2)
ans =
   -0.5000
```

If your output does not match ours, check your function code for errors. If you still have difficulty, revisit **Trouble Shooting Function M-files** in the section "Simple Function M-files" of Chapter 3. The following code[2] should produce an image similar to that shown in Figure 7.1.

```
>> [t,x]=ode45('test',[0,2*pi],3);
>> plot(t,x)
>> title('The solution of x''=cos(t)/(2x-2), with x(0)=3.')
>> xlabel('t')
>> ylabel('x')
>> grid
```

---

[1] If it bothers you that the function M-file name `test` does not match the function name f in the differential equation $x' = f(t, x)$, simply rewrite the function, changing the first line to `function xprime=f(t,x)` and saving the file as `f.m`. Thereafter, anywhere we reference the file `test`, simply replace this with a reference to `f`.

[2] It is instructive to leave the semicolon off the statement `[t,x]=ode45('test',[0,2*pi],3)` in order to see the output from the `ode45` routine. Each row in the solution vector x corresponds to a time that can be found in the same row of the time vector t. You might want to enter `[t,x]` or `[t(1:20),x(1:20)]` at the MATLAB prompt to see this connection.

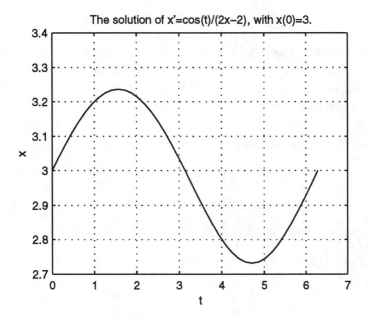

Figure 7.1. The solution of initial value problem (7.2)

Systems of First Order Equations

Actually, systems are no harder to handle using `ode45` than are single equations. Consider the following system of n first order differential equations:

$$
\begin{aligned}
x_1' &= f_1(t, x_1, x_2, \cdots, x_n), \\
x_2' &= f_2(t, x_1, x_2, \cdots, x_n), \\
&\vdots \\
x_n' &= f_n(t, x_1, x_2, \cdots, x_n).
\end{aligned}
\tag{7.4}
$$

System (7.4) can be written in vector form as follows.

$$
\begin{bmatrix} x_1 \\ x_2 \\ \vdots \\ x_n \end{bmatrix}'
=
\begin{bmatrix}
f_1(t, [x_1, x_2, \cdots, x_n]^T) \\
f_2(t, [x_1, x_2, \cdots, x_n]^T) \\
\vdots \\
f_n(t, [x_1, x_2, \cdots, x_n]^T)
\end{bmatrix}
\tag{7.5}
$$

Now, here comes the key to the whole business of using MATLAB's ODE suite. If we set the vector $\mathbf{x} = [x_1, x_2, \ldots, x_n]^T$, then system (7.5) becomes

$$
\mathbf{x}' =
\begin{bmatrix}
f_1(t, \mathbf{x}) \\
f_2(t, \mathbf{x}) \\
\vdots \\
f_n(t, \mathbf{x})
\end{bmatrix}.
\tag{7.6}
$$

103

Finally, if we define $\mathbf{F}(t, \mathbf{x}) = [f_1(t, \mathbf{x}), f_2(t, \mathbf{x}), \ldots, f_n(t, \mathbf{x})]^T$, then system (7.6) can be written as

$$\mathbf{x}' = \mathbf{F}(t, \mathbf{x}), \tag{7.7}$$

which, most importantly, has a form identical to the single first order differential equation, $x' = f(t, x)$, used in Example 1. Consequently, if `function xprime=F(t,x)` is the first line of a function ODE file, it is extremely important that you understand that x is a vector[3] with entries `x(1)`, `x(2)`,..., `x(n)`. Confused? Perhaps a few examples will clear the waters.

Example 2. *Use* `ode45` *to solve the initial value problem*

$$
\begin{aligned}
x_1' &= x_2 - x_1^2, \\
x_2' &= -x_1 - 2x_1 x_2,
\end{aligned}
\tag{7.8}
$$

on the interval [0, 10], *with initial conditions* $x_1(0) = 0$ *and* $x_2(0) = 1$.

We can write system (7.8) as a vector equation.

$$
\begin{bmatrix} x_1 \\ x_2 \end{bmatrix}' = \begin{bmatrix} x_2 - x_1^2 \\ -x_1 - 2x_1 x_2 \end{bmatrix}
\tag{7.9}
$$

If we set $\mathbf{F}(t, [x_1, x_2]^T) = [x_2 - x_1^2, -x_1 - 2x_1 x_2]^T$ and $\mathbf{x} = [x_1, x_2]^T$, then system (7.9) takes the form $\mathbf{x}' = \mathbf{F}(t, \mathbf{x})$. The key thing to realize is the fact that $\mathbf{x} = [x_1, x_2]^T$ is a vector with two components, x_1 and x_2. Similarly, $\mathbf{x}' = [x_1', x_2']^T$ is also a vector with two components, x_1' and x_2'.

Open your editor and create the following ODE file[4].

```
function xprime=F(t,x)

xprime=zeros(2,1);    %The output must be a column vector
xprime(1)=x(2)-x(1)^2;
xprime(2)=-x(1)-2*x(1)*x(2);
```

Save the file[5] as F.m. The line `xprime=zeros(2,1)` *initializes* `xprime`, creating a *column* vector with two rows and 1 column. Each of its entries is a zero. Note that the next two lines of the ODE file duplicate exactly what you see in system (7.8).

Even though we write $\mathbf{x} = [x_1, x_2]^T$ in the narrative, it is extremely important that you understand that x is a *column* vector; that is,

$$\mathbf{x} = \begin{bmatrix} x_1 \\ x_2 \end{bmatrix}.$$

---

[3] In MATLAB, if `x=[2;4;6;8]`, then `x(1)=2`, `x(2)=4`, `x(3)=6`, and `x(4)=8`.

[4] Even though this system is autonomous, the ODE suite does not permit you to write the first line of the ODE file as follows: `function xprime=F(x)`. If you did, you would receive lots of error messages.

[5] In this example, we've matched the function M-file name, F, with the function name in the differential equation $\mathbf{x}' = \mathbf{F}(t, \mathbf{x})$. Of course, you are free to give your function M-file a distinctive name, such as `function zprime=test2(t,z)`. You would save this file as `test2.m` and call `ode45` in the following manner: `[t,z]=ode45('test2',[0,10],[0;1])`.

Therefore, the variable x in the ODE file is also a *column* vector.

As always, it is an excellent idea to test that your function ODE file is working properly before continuing. Because $\mathbf{F}(t, [x_1, x_2]^T) = [x_2 - x_1^2, -x_1 - 2x_1x_2]^T$, we have that $\mathbf{F}(2, [3, 4]^T) = [4 - 3^2, -3 - 2(3)(4)]^T = [-5, -27]^T$. Test this result at the MATLAB prompt.

```
>> F(2,[3;4])
ans =
    -5
   -27
```

Do you see how the function F(t,x) expects two inputs, a scalar and a column vector? Because the time is $t = 2$, we used t=2. Further, since $x = [x_1, x_2]^T = [3, 4]^T$, we input x=[3;4], a column vector whose first entry is x(1)=3 and whose second entry is x(2)=4. Because this system is autonomous and independent of t, the command F(5,[3;4]) should also produce [-5;-27]. Try it!

We are almost ready to call ode45. But first, we must establish our initial condition. Recall that the initial conditions for system (7.8) were given as $x_1(0) = 0$ and $x_2(0) = 1$. Consequently, our initial condition vector will be

$$\mathbf{x}(0) = \begin{bmatrix} x_1(0) \\ x_2(0) \end{bmatrix} = \begin{bmatrix} 0 \\ 1 \end{bmatrix}.$$

Of course, this is a column vector, so one would use x0=[0;1] in a call to ode45.

```
>> [t,x]=ode45('F',[0,10],[0;1]);
```

Typing whos at the MATLAB prompt will reveal information about the variables in your workspace[6].

```
>> whos
  Name       Size          Bytes  Class

  t          89x1            712  double array
  x          89x2           1424  double array
```

Note that t and x have an identical number of rows, as they should. Each row in the solution array x corresponds to a time that can be found in the same row of the column vector t. The vector x has two columns. The first column of x contains the solution for x_1, while the second column of x contains the solution x_2. Typing [t,x] at the MATLAB prompt and viewing the resulting output will help make this connection[7].

The MATLAB command plot(v,A) will produce a plot of each column of the matrix A versus the vector v. Consequently, the command plot(t,x) should produce a plot of each column of the matrix x versus the vector t. The following commands were used to produce the plots of x_1 versus t and x_2 versus t in Figure 7.2.

---

[6] You may have more variables in your workspace than the number shown here.

[7] For example, note that the time values in the vector t run from 0 through 10, as they should.

```
>> close all
>> plot(t,x)
>> title('x_1''=x_2-x_1^2 and x_2''=-x_1-2x_1x_2')
>> xlabel('t')
>> ylabel('x_1 and x_2')
>> legend('x_1','x_2')
>> grid
```

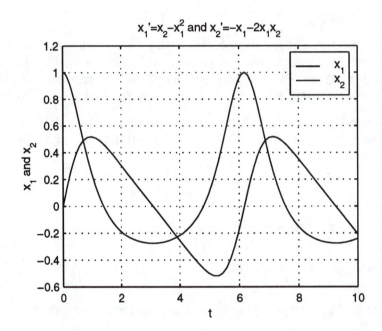

Figure 7.2. Solution of system (7.8) using `ode45`

If only the first component of the solution is wanted, enter `plot(t,x(:,1))`. The colon in the notation `x(:,1)` indicates[8] that we want all rows, and the 1 indicates that we want the first column. Similarly, if only the second component is wanted, enter `plot(t,x(:,2))`. This is an example of the very sophisticated indexing options available in MATLAB.

It is also possible to plot the components against each other with the commands

```
>> plot(x(:,1),x(:,2))
>> grid
>> title('x_1''=x_2-x_1^2 and x_2''=-x_1-2x_1x_2')
>> xlabel('x_1')
>> ylabel('x_2')
```

The result is called a *phase plane* plot, and it is shown in Figure 7.3. Readers may also find the command sequence `close all`, `comet(x(:,1),x(:,2))` both entertaining and informative.

---

[8] Some MATLAB users pronounce the notation `x(:,1)` as "x, every row, first column."

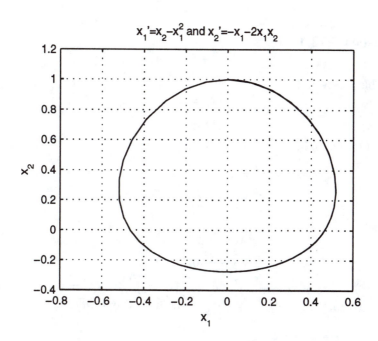

Figure 7.3. the phase plane solution of system (7.8)

Another way to present the solution to the system graphically is in a three dimensional plot, where both components of the solution are plotted as separate variables against t. MATLAB does this using the command `plot3`. For example, enter `plot3(t,x(:,1),x(:,2))` to see the plot with t along the x-axis, and the two components of the solution along the y-axis and z-axis, respectively. Alternatively, enter `plot3(x(:,1),x(:,2),t)` to see the solution with t along the z-axis, and the two components of the solution along the x-axis and y-axis, respectively. Type `rotate3d`, then use your mouse to click and drag the plot to rotate the view. Readers will also want to try the command sequence `close all`, `comet3(x(:,1),x(:,2),t)`.

Second Order Differential Equations

To solve a single second order differential equation it is necessary to replace it with the equivalent first order system. For the equation

$$y'' = f(t, y, y'), \tag{7.10}$$

we set $x_1 = y$, and $x_2 = y'$. Then $\mathbf{x} = (x_1, x_2)$ is a solution to the first order system

$$\begin{aligned} x_1' &= x_2, \\ x_2' &= f(t, x_1, x_2). \end{aligned} \tag{7.11}$$

Conversely, if $\mathbf{x} = [x_1, x_2]^T$ is a solution of the system in (7.11), we set $y = x_1$. Then we have $y' = x_1' = x_2$, and $y'' = x_2' = f(t, x_1, x_2) = f(t, y, y')$. Hence y is a solution of the equation in (7.10).

Example 3. *Plot the solution of the initial value problem*

$$y'' + yy' + y = 0, \quad y(0) = 0, \ y'(0) = 1, \tag{7.12}$$

107

on the interval [0, 10].

First, solve equation (7.12) for y''.

$$y'' = -yy' - y \tag{7.13}$$

Introduce new variables for y and y'.

$$x_1 = y \tag{7.14}$$
$$x_2 = y'$$

Then we have by (7.13) and (7.14),

$$x_1' = y' = x_2, \tag{7.15}$$
$$x_2' = y'' = -yy' - y = -x_1 x_2 - x_1,$$

or, more simply,

$$x_1' = x_2, \tag{7.16}$$
$$x_2' = -x_1 x_2 - x_1.$$

If we let $\mathbf{F}(t, [x_1, x_2]^T) = [x_2, -x_1 x_2 - x_1]^T$ and $\mathbf{x} = [x_1, x_2]^T$, then $\mathbf{x}' = [x_1', x_2']^T$ and system (7.16) takes the form $\mathbf{x}' = \mathbf{F}(t, \mathbf{x})$, prompting the writing of the following ODE file.

```
function xprime=example3(t,x)
xprime=zeros(2,1);
xprime(1)=x(2);
xprime(2)=-x(1)*x(2)-x(1);
```

Save the file as `example3.m`. Note that $\mathbf{F}(1, [2, 3]^T) = [3, -(2)(3) - 2]^T = [3, -8]^T$ and test your function as follows:

```
>> example3(1,[2;3])
ans =
        3
       -8
```

Re-examine (7.12), establish the initial condition (remember, $x_1 = y$, $x_2 = y'$)

$$\mathbf{x}(0) = \begin{bmatrix} x_1(0) \\ x_2(0) \end{bmatrix} = \begin{bmatrix} y(0) \\ y'(0) \end{bmatrix} = \begin{bmatrix} 0 \\ 1 \end{bmatrix},$$

and call `ode45`.

```
>> [t,x]=ode45('example3',[0,10],[0;1]);
```

It is important to note that the original question called for a solution of $y'' + yy' + y = 0$. In (7.14), recall that y equals x_1. Consequently, a plot of y versus t, or x_1 versus t, is required. The commands

```
>> plot(t,x(:,1))
>> title('y''''+yy''+y=0, y(0)=0, y''(0)=1')
>> xlabel('t')
>> ylabel('y')
>> grid
```

should produce an image similar to that in Figure 7.4.

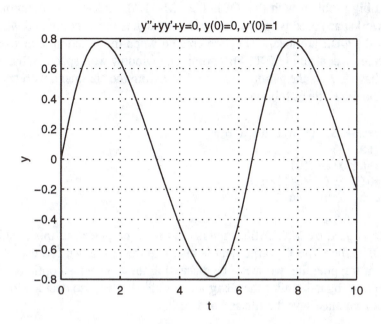

Figure 7.4. The solution of initial value problem (7.14)

The Lorenz System

The solvers in MATLAB can solve first order systems containing as many equations as you like. As an example we will solve the Lorenz system. This is a system of three equations which was published in 1963 by the meteorologist and mathematician E. N. Lorenz. It represents a simplified model for atmospheric turbulence beneath a thunderhead. The equations are

$$
\begin{aligned}
x' &= -ax + ay, \\
y' &= rx - y - xz, \\
z' &= -bz + xy,
\end{aligned}
\qquad (7.17)
$$

where a, b, and r are positive constants.

We could proceed with the variables x, y, and z, but to make things a bit simpler, set $u_1 = x$, $u_2 = y$, and $u_3 = z$. With these substitutions, system (7.17) takes the following vector form.

$$
\begin{bmatrix} u_1 \\ u_2 \\ u_3 \end{bmatrix}' = \begin{bmatrix} -au_1 + au_2 \\ ru_1 - u_2 - u_1u_3 \\ -bu_3 + u_1u_2 \end{bmatrix}
\qquad (7.18)
$$

If we let $\mathbf{F}(t, [u_1, u_2, u_3]^T) = [-au_1 + au_2, ru_1 - u_2 - u_1u_3, -bu_3 + u_1u_2]^T$ and $\mathbf{u} = [u_1, u_2, u_3]^T$, then $\mathbf{u}' = [u_1', u_2', u_3']^T$ and system (7.18) takes the form $\mathbf{u}' = \mathbf{F}(t, \mathbf{u})$, prompting one to write the ODE file as follows.

```
function uprime=F(t,u)
uprime=zeros(3,1);
uprime(1)=-a*u(1)+a*u(2);
uprime(2)=r*u(1)-u(2)-u(1)*u(3);
uprime(3)=-b*u(3)+u(1)*u(2);
```

However, there is a big problem with this ODE file. MATLAB executes its functions in a workspace different from the workspace used when you execute commands in the command window. This means that all variables in the ODE are local. They are created when the function is invoked, then forgotten when the function returns control to MATLAB's command window workspace. What does this mean? It means that we will have to *pass* the parameters a, b, and r from the command workspace to the function ODE file. Rewrite the ODE file as follows.

```
function uprime=lor1(t,u,flag,a,b,r)
uprime=zeros(3,1);
uprime(1)=-a*u(1)+a*u(2);
uprime(2)=r*u(1)-u(2)-u(1)*u(3);
uprime(3)=-b*u(3)+u(1)*u(2);
```

The argument `flag` is used by MATLAB's solvers, and must be placed in this position when passing parameters to the ODE file. The flag argument is also required for advanced use of MATLAB's ODE solvers, such as event trapping, but we won't be discussing these advanced options. Note that we have placed our parameters a, b, and c after the `flag` argument[9]. We've also decided to name our function lor1 instead of F, so we must save the file as `lor1.m`[10].

We will use $a = 10$, $b = 8/3$, and $r = 28$ and the initial condition

$$\mathbf{u}(0) = \begin{bmatrix} u_1(0) \\ u_2(0) \\ u_3(0) \end{bmatrix} = \begin{bmatrix} 1 \\ 2 \\ 3 \end{bmatrix}.$$

The commands

```
>> [t,u]=ode45('lor1',[0,7],[1;2;3],[],10,8/3,28);
>> plot(t,u)
>> grid
>> title('A solution to the Lorenz system')
>> xlabel('t')
>> legend('u_1','u_2','u_3')
```

should produce an image similar to that in Figure 7.5.

Global Variables. If you find the passing of variables through arguments in the function M-file confusing, perhaps you would find the use of global variables a bit easier to understand. First change the function ODE file, `lor1.m`, as follows:

---

[9] Although the Lorenz system has only three parameters, you can use more than three parameters should your problem call for it. Just add the extra parameters after the a, b, and r. Fortunately, mathematicians and scientists like to scale variables, thereby introducing dimensionless variables and effectively eliminating a number of parameters. We used this technique on the pendulum problem in Chapter 6.

[10] It would be natural to name this file `lorenz.m`, but there is already an M-file in the MATLAB directory tree with that name. If you enter `lorenz` at the MATLAB prompt you will see a solution to the Lorenz system (7.17) displayed in a very attractive manner.

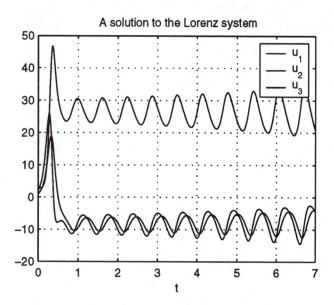

Figure 7.5. A solution to the Lorenz system.

```
function uprime=lor2(t,u)
global A B R
uprime=zeros(3,1);
uprime(1)=-A*u(1)+A*u(2);
uprime(2)=R*u(1)-u(2)-u(1)*u(3);
uprime(3)=-B*u(3)+u(1)*u(2);
```

Save this second version as `lor2.m`. Note that we always follow the practice of using uppercase names for global variables.

Next, declare the variables global in the MATLAB command window workspace, then initialize the variables.

```
>> global A B R
>> A=10;
>> B=8/3;
>> R=28;
```

The global variables A, B, and R are now available in both your command window workspace and in the function workspace created when your ODE file is called. You can now find the numerical solution of the Lorenz system with the command

```
>> [t,u]=ode45('lor2',[0,7],[1;2;3]);
```

Eliminating Transient Behavior. In Figure 7.5, there appears to be transient behavior in the interval $0 \le t \le 1$. Let's compute the solution over a longer period, say $0 \le t \le 100$. (If your computer is slow, use the upper bound of 20 instead of 100.) We will then plot the part corresponding to $t > 1$ in

111

three dimensions. This is easily accomplished. If the global variables A, B, and R are still declared and initialized in your command window workspace, then you can use the second version of the Lorenz ODE file, lor2.m, as follows:

```
>> [t,u]=ode45('lor2',[0,100],[1;2;3]);
>> N=find(t>1);
>> v=u(N,:);
>> plot3(v(:,1),v(:,2),v(:,3))
>> grid
>> title('The Lorenz attractor')
>> xlabel('x')
>> ylabel('y')
>> zlabel('z')
>> rotate3d
```

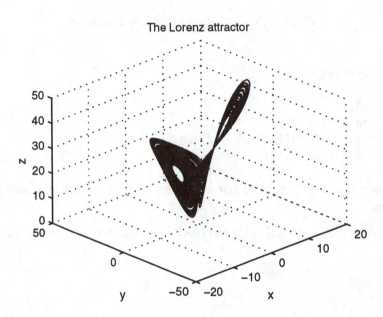

Figure 7.6. The Lorenz attractor.

Some explanation is necessary. First, the command N= find(t>1); produces a list of the indices of those elements of t which satisfy $t > 1$. Then v=u(N,:); produces a matrix containing the rows of u with indices in N. The result is that v consists of only those solution points corresponding to $t > 1$, and the resulting plot represents the long time behaviour of the solution after the transient effects are gone.

After a little grooming we get Figure 7.6, a fine picture of the Lorenz attractor. The Lorenz system, with this particular choice of parameters, has the property that any solution, no matter what its initial point, is attracted to this rather complicated, butterfly-shaped, set. We will return to this in the exercises. There we will also examine the Lorenz system for different values of the parameters.

Finally, the `rotate3d` command enables rotation with the mouse. Click and drag the mouse in the on the axes in Figure 7.6 to rotate the axes to a different view of the Lorenz attractor. If the animation is unbearably slow, go back and repeat the commands for the construction of the Lorenz attractor, only use a smaller time interval, say [0, 20].

For an animated plot, try

```
>> close all, comet3(v(:,1),v(:,2),v(:,3)).
```

Improving Accuracy

It has been mentioned before that `ode45` chooses its own step size in order to achieve a predetermined level of accuracy. Although the default level is sufficient for many problems, sometimes it will be necessary to improve that "predetermined level of accuracy."

The accuracy is controlled by the user by defining two optional inputs to `ode45`, and the other solvers. These are the *relative tolerance* and the *absolute tolerance*. If y^k is the computed solution at step k, then each component of the solution is required to satisfy its own error restriction. This means that we consider an estimated error vector, which has a component for every component of y, and it is required that each component of the error vector satisfy

$$|\text{estimated error}_j^k| \leq \max(|y_j^k| \times \text{RelTol}, \text{AbsTol}_j). \tag{7.19}$$

Here `RelTol` is the *relative tolerance*, and `AbsTol` is the *absolute tolerance*. Notice that the relative tolerance is a number, but the absolute tolerance is a vector quantity, with a component for each equation in the system being solved. This allows the user to set the tolerance differently for different components of the solution. The default values for `RelTol` is 10^{-3}, and `AbsTol` is a vector, each component of which is 10^{-6}.

Choosing the Tolerances. The two tolerances provide the user with a wide variety of potential error controlling strategies. For example, it is possible to set the relative tolerance to 0, and then to use only the absolute tolerance to control the errors. It is also possible to set the absolute tolerance to 0, and use only the relative tolerance.

However the philosophy behind the requirement in (7.19), is that the relative tolerance is chosen so that the estimated error is bounded by certain fraction of the size of the computed quantity y_j^k for most values of that quantity. The absolute tolerance should come into play only for small values of y_j^k. This is to prevent the routine from trying too hard if the computed quantity turns out to be very small at some steps. Thus the components of the absolute tolerance should be chosen to be pretty small in comparison to the expected values of the computed quantity.

Let's talk about choosing the relative tolerance first. This is the easier case since we are simply talking about the number of significant digits we want in the answer. For example, the default relative tolerance is 10^{-3}, or 0.1%. In an ideal world this will ensure that the first three digits of the computed answer will be correct. If we want five significant digits we should choose 10^{-5}, or 0.001%.

It would be nice if things were as simple as these considerations make them seem. Unfortunately the inequalities in (7.19) at best control the errors made at any individual step. These errors propagate, and errors made at an early step can propagate to something much larger. In fact, the errors can propagate

113

exponentially. The upshot is that over long time periods computational solutions can develop serious inaccuracies. In a situation like this it is a good idea to have a way of checking your solution.

It is usually a good idea is to introduce a safety factor into your choice of tolerances. For example if you really want three significant digits, it might be good to choose the relative tolerance smaller than 10^{-3}, say 10^{-5}. Even then it is imperative that you check your answer with what you expect and reduce the tolerance even further if it seems to be called for.

To choose the absolute tolerance vector, remember that the philosophy is that the absolute tolerance will dominate in (7.19) only when the computed value y_j^k is small relative to the normal values of that component multiplied by the relative tolerance. One way to do this is to set

$$\text{AbsTol}_j = 10^{-3} \times \text{normal value of } |y_j^k| \times \text{RelTol}. \tag{7.20}$$

Of course that 10^{-3} can be replaced by a smaller or larger number if the situation at hand merits it.

Let's use these considerations to decide what tolerances to use for the Lorenz equations. Since we are only going to plot them, two place accuracy should be sufficient. Thus, the default relative tolerance of 10^{-3} would seem to be sufficient. If we were to compute over longer periods of time, we might want to choose a smaller value, but without examining the computed results a relative tolerance smaller than 10^{-5} does not seem justified.

To choose the absolute tolerance vector, we notice from Figures 7.5 and 7.6 that the normal ranges of the components are $|x| \leq 20$, $|y| \leq 30$, and $|z| \leq 40$. Hence according to (7.20), we should choose

$$\text{AbsTol}_1 \leq 10^{-3} \times 20 \times 10^{-3} = 2 \times 10^{-5},$$
$$\text{AbsTol}_2 \leq 10^{-3} \times 30 \times 10^{-3} = 3 \times 10^{-5},$$
$$\text{AbsTol}_3 \leq 10^{-3} \times 40 \times 10^{-3} = 4 \times 10^{-5}.$$

Again we see that the default tolerance of 10^{-6} for each seems to be adequate.

As we have approached the issue of choosing tolerances everything depends on choosing the relative tolerance. From our earlier discussion this seems simple enough. However, there are situations in which more accuracy may be required than appears on the surface. Suppose, for example, that it is the difference of two components of the solution which is really important. To be more specific, suppose we compute a solution $\mathbf{y}$, and what we are really interested in is $y_1 - y_2$. Suppose in addition that y_1 and y_2 are very close to each other. Suppose in fact that we know they will always agree in the first 6 digits. If we set the relative tolerance to 10^{-7}, then in an ideal world we will get 7 significant digits for y_1 and y_2. For example we might get $y_1 = 1234569 * * * **$ and $y_2 = 1234563 * * * **$, where the asterisks indicate digits that are not known accurately. Then $y_1 - y_2 = 6 * * * **$, so we only know 1 significant digit in the quantity we are really interested in. Consequently to get more significant digits in $y_1 - y_2$ we have to choose a smaller relative tolerance. To get 3 significant digits we need 10^{-9}, or with a suitable safety factor, 10^{-11}.

Using the Tolerances in the Solvers. Having chosen our tolerances, how do we use them in `ode45` and the other MATLAB solvers? These solvers have many options, many more that we will describe in this book. All of the options are input into the solver using an options vector. Suppose, for example, that we wanted to solve the Lorenz system with relative tolerance 10^{-5} and absolute tolerance vector $[10^{-8}, 10^{-7}, 10^{-7}]$. We build the options vector with the command `odeset`, and then enter it as a

114

parameter to `ode45`. If we use the first version of the Lorenz ODE file, `lor1.m`, then we proceed as follows:

```
>> options = odeset('RelTol',1e-5,'AbsTol',[1e-8 1e-7 1e-7]);
>> [t,u]=ode45('lor1',[0,100],[1;2;3],options,10,8/3,28);
```

If the run takes inordinately long, try a shorter time span, say `[0,20]` instead of `[0,100]`. Notice that the syntax used with `ode45` is exactly the same as was used earlier for the Lorenz system, except that we have inserted the parameter `options` between the initial conditions and the parameters sent to the routine.

If you wish to use global variables, then use the second version of the Lorenz ODE file with

```
>> options = odeset('RelTol',1e-5,'AbsTol',[1e-8 1e-7 1e-7]);
>> [t,u]=ode45('lor2',[0,100],[1;2;3],options);
```

Again, this is assuming that you still have your global declarations in place and initialized.

A Cautionary Example. Consider the initial value problem

$$x' = t(x - 1), \quad x(-a) = 0. \tag{7.21}$$

You will notice that this is a linear equation, and we can easily find the exact solution by separating the variables,

$$x(t) = 1 - e^{(t^2 - a^2)/2}.$$

We will be interested in this initial value problem when a is a number between 0 and 10.

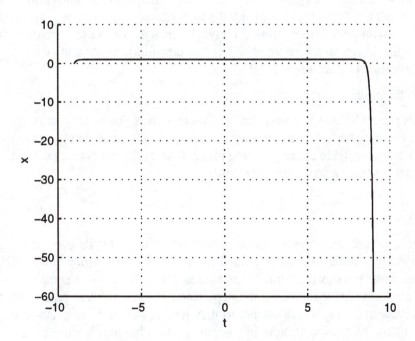

Figure 7.7. The computed solution to the initial value problem (7.21) when $a = 9$.

115

We will be looking at the solution graphically, so we will need only 2 significant figures in the solution. This leads us to believe that the default tolerances should be suitable for our calculations. The reader is encouraged to look at the solution computed over the interval $[-a, a]$ for a number of values of a using the default tolerances.

It is easy to come to a crude judgement about the accuracy by simply looking at the graph of the solution. The exact solution is an even function of t, and in particular $x(a) = x(-a) = 0$. By doing the computation and plotting the solution, you will notice that when a is as small as 5, the computation with the default tolerances is visibly not accurate.

But let's cut right to the point and use 10^{-13} for the relative tolerance and 10^{-16} for the absolute tolerance. If we compute the solution for $a = 9$ we get the result shown in Figure 7.7. Clearly the computed value of the solution at $t = 9$ is nowhere near 0, so the computed result is incorrect.

The initial value problem in (7.21) is seemingly very nice, yet computationally it is a nightmare. It is not difficult to discover why. Notice first that $x(t) = 1$ is another solution to the differential equation. Then look at the exact solution $x(t) = 1 - e^{(t^2-a^2)/2}$. When $t = 0$, we have $x(0) = 1 - e^{-a^2/2}$. If $a = 9$, this becomes $x(0) = 1 - 2.58 \times 10^{-18}$, which is equal to $y(0) = 1$ to 17 significant figures. MATLAB computes accurately only to 16 significant figures, so MATLAB cannot distinguish between $x(0)$ and $y(0)$. This means that MATLAB cannot distinguish between these two solutions and many others. The solver is continually making small mistakes. A small mistake when the solution is near the point $(0,1)$ will take the solution to another which ends up quite a distance away when t reaches 9.

The earlier considerations were for solutions over "relatively short" intervals of the independent variable. When solutions are required over relatively long intervals, we have a whole new ball game. No solver can give reliable results over long intervals for all equations. As in this example, many equations have regions where solutions are extremely sensitive to initial conditions. This means that two solutions which are very close can eventually move very far apart. From a computational point of view, it must be recognized that small errors are being made all the time. Even if the error is very small, in a region of extreme sensitivity to initial conditions the error might be enough to move to a solution which is diverging from the one sought. This is in the nature of differential equations, and cannot be anticipated by a solver without the intervention of the user.

So, what is the moral of this tale?

No solution should be accepted uncritically. Compare the solution to what is expected. If nothing else is possible, compute the solution again with a decreased tolerance, and compare the two solutions. If they differ, then unacceptable errors are being made. Reduce the tolerances further and try again until you are satisfied that you ae getting consistent answers.

Kinky Plots

Sometimes a variable step solver can be almost too accurate for its own good. It will use such long step sizes that when the solution is plotted the result is a very kinky curve. The latest version of `ode45` almost never has this problem. It can take as few as 10 steps to compute a solution, but the routine interpolates between the solution at the step points. The interpolation process uses the differential equation to compute the interpolating points, so they remain almost as accurate as the solution at the step points. Furthermore, the use of more interpolating points has almost no cost in terms of increased computational time.

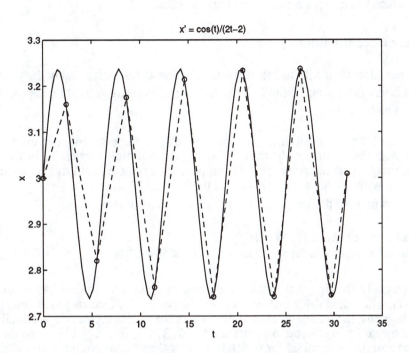

Figure 7.8. The effect of the refine option.

The number of interpolating points is controlled by the option `Refine`. The default value of `Refine` is 4. This should be sufficient to insure that most solution curves are smooth. If it happens that a solution curve is not smooth enough, the solution can be recomputed with a larger value of `Refine`. As an example, consider the first equation in this chapter, $x' = \cos(t)/(2x - 2)$. This time we solve it over the interval $[0, 10\pi]$. The solution is periodic with period 2π, so we get five periods. One would expect the graph of such a function to be kinky, unless a large number of points were plotted. Indeed, the default value of `refine` does lead to a somewhat kinky graph of the solution. So, let's set `Refine = 8`. The following list of commands

```
>> options = odeset('refine',8);
>> [t,x] = ode45('test',[0,10*pi],3,options);
>> L=(length(t)-1)/8;       % The number of steps.
>> tp=t(8*[0:L]+1);     % The values of t at the step points.
>> xp=x(8*[0:L]+1);       % The values of x at the step points.
>> plot(t,x,tp,xp,'o',tp,xp,'--')
```

results in Figure 7.8. In this figure, the solid curve is the plot of the entire solution, and the circles represent the solution at the step points. Consequently, the dashed curve is what we would get without the refine option.

Notice that we once more used the command `odeset`. In this case as `odeset('refine',8)`. The effect of the command is to set up an options vector for `ode45`, with the refine option set to 8. If we had wanted to use different tolerances as well, we could have used something like

```
>> options = odeset('refine',8,'reltol',1e-5,'abstol',1e-8);
```

117

The order in which the choices are entered is unimportant.

Behavior Near Discontinuities

Any numerical method will run into difficulties where the equation and/or the solution to the equation has a point of discontinuity, and ode45 is no exception. There are at least three possible outcomes when ode45 meets a discontinuity.

1. ode45 can integrate right through a discontinuity in the equation, not even realizing it is there. In this case the accuracy of the result is highly doubtful, especially in that range beyond the discontinuity. This phenomenon will happen, for example, with the initial value problem $x' = x/(t-1)$, with initial value $x(0) = 1$ on the interval $[0, 2]$.

2. ode45 can find the discontinuity and report it with a comment like

```
Step failure at 1.129671e+01
      with a minimum step size of 1.003350e-13
```

This is usually the sign that there is a discontinuity in the solution. An example of this is $x' = x/(1-t)$, with initial value $x(0) = 1$ on the interval $[0, 1]$. One very nice thing about ode45 is that if this happens, the output up to the point where the routine stops is made available. For example, if you execute [t,w] = ode45('gizmo',[0,3.4],[0; 1; 2]);, and the computation stops because a step size is called for which is smaller than the minimum allowed, then the variables t and w will contain the results of the computation up to that point.

3. ode45 can choose smaller and smaller step sizes upon the approach to a discontinuity and go on calculating for a very long time — hours in some cases. For cases like this, it is important for the user to know how to stop a calculation in MATLAB. On most computers the combination of the control key and C depressed simultaneously will do the trick.

Stiff Equations

Solutions to differential equations often have components which are varying at different rates. If these rates differ by a couple of orders of magnitude, the equations are called stiff. Stiff equations will not arise very often in this manual, but we will say a word about them in case the reader should have to deal with one. We will also put some examples in the exercises.

An example of a stiff system is the van der Pol equation

$$x' = y,$$
$$y' = -x + \mu(1 - x^2)y,$$

when the parameter μ is very large, say $\mu = 1000$.

Runge-Kutta solution methods do not work well with stiff equations. The fast varying component requires a Runge-Kutta algorithm to use very small steps — so small that it takes a prohibitively long time to compute the solution. The new suite of ode solvers contains four routines, ode15s, ode23s, ode23t, and ode23tb, which are designed to solve stiff equations. ode15s is the first of these to try.

A very attractive feature of the new suite of solvers is that the calling syntax is the same for all of them. Thus, the syntax for using ode15s is the same as we have been using with ode45. For example,

```
[t,x] = ode15s('vdpol',tspan,init);
```

will compute the solution of the van der Pol system, if the system is described in the function M-file `vdpol.m`, `tspan = [t0,tf]` is the interval of computation, and `init = [x0;y0]` is a column vector of initial conditions. These routines also take an options vector as an additional parameter, and the makeup of that is again the same as it is for `ode45`. In particular the same options are available. There is one change — the default value of `Refine` is 1.

But how do we tell if a system is stiff? It is often obvious from the physical situation being modeled that there are components of the solution which vary at rates that are significantly different, and therefore the system is stiff. Sometimes the presence of a large parameter in the equations is a tip off that the system is stiff. However, there is no general rule that allows us to recognize stiff systems. A good operational rule is to try to solve using `ode45`. If that fails, or is very slow, try `ode15s`.

Other Possibilities

The MATLAB ODE Suite consists of a very powerful group of solvers. We have only touched the surface of the options available to control them, and the variations of the ways in which they can be used. To mention just a few, it is possible to modify the output of the solver so that the solution is automatically plotted in the way the user desires. It is also possible to get the solvers to detect events. For example the 3-body problem of the sun, earth, and moon can be modeled by a system of 12 ODEs, and the solver can be set up to automatically detect the times of eclipses of the sun and moon.

There is a very nice description of the ODE Suite in *Using MATLAB*, which is part of the MATLAB documentation. You should be able to find it by entering `helpdesk` at the MATLAB prompt. After the MATLAB Help Desk opens, click on Onlines Manuals (in PDF). You will see "Using MATLAB" listed in the next list.

There is also available a more technical paper dealing the the ODE Suite, by Larry Shampine and Mark Reichelt, who built the Suite. It is also available from the Onlines Manuals list, where it says MATLAB ODE Suite.

Exercises

1. MATLAB's solvers are able to handle multiple initial conditions. This is a relatively unknown fact and isn't even documented in the MATLAB help files. However, it does work, *provided you make your function ODE file array smart*. For example, create the function

   ```
   function yprime=steady(t,y)
   yprime=-2*y+2*cos(t).*sin(2*t);
   ```

   and save the file as `steady.m`.

   a) To solve the equation $y' = -2y + 2\cos t \sin 2t$, for initial conditions $y(0) = -5$, $y(0) = -4,\ldots$, $y(0) = 5$, enter the code `[t,y]=ode45('steady',[0,30],-5:5);`.

   b) Graph the solutions with `plot(t,y)` and note that each solution approaches a periodic *steady-state* solution.

   c) Repeat parts (a) and (b) with time spans $[0, 10]$ and $[0, 2\pi]$ to get a closer look at the convergence to the steady-state solution.

2. Another nice feature of MATLAB's solvers is the ability to find solutions at specific times. This is useful when you want to compare two functions at specific time points or place the output from multiple calls to ode45 in a matrix, where it is required that columns and rows have equal length. Enter `[t,y]=ode45('steady', 0:.25:3,1);`, then enter `[t,y]`. This should clarify what is going on.

119

3. Another little known fact is MATLAB's solvers can automatically plot solutions *as they are computed*. All you need to do is leave off the output variables when calling the solver. Try `ode45('steady',[0,30],-5:5)`.

4. Plot solutions for each of the initial values given in the following differential equations.

   a) $y' + 4y = 2\cos t + \sin 4t$, $y(0) = -5, -4, \ldots, 5$

   b) $y' + 4y = t^2$, $y(0) = -9, -8, -7, -1, 1, 7, 8, 9$

5. Separate variables and find an explicit solution of $x' = t/(1+x)$, $x(0) = 1$.

   a) Use `ode45` to find the numerical solution of the initial value problem on the interval [0, 3]. Store the solution in the variables `t` and `x_ode45`.

   b) Use the explicit solution to find exact solution values at each time instance in the variable `t`. Store the results in the variable `x_exact`. Compare the solutions with `plot(t,x_ode45,t,x_exact,'r*')`. Obtain a printout with a legend.

6. Write a function ODE file for each of the following initial value problems. Provide plots of x_1 versus t, x_2 versus t, and x_2 versus x_1 on the time interval provided. See Exercise 1 in Chapter 6 for a nice method you can use to arrange these plots.

   a) $x_1' = x_2$
   $x_2' = (1 - x_1^2)x_2 - x_1$
   $x_1(0) = 0, x_2(0) = 4, [0, 10]$

   b) $x_1' = x_2$
   $x_2' = -25x_1 + 2\sin 4t$
   $x_1(0) = 0, x_2(0) = 2, [0, 2\pi]$

7. Each of MATLAB's solvers' built-in plotting routines work equally well with systems. For example, create

   ```
   function yprime=heart(t,y)
   yprime=zeros(2,1);
   yprime(1)=y(2);
   yprime(2)=-16*y(1)+4*sin(2*t);
   ```

   and save as `heart.m`.

   a) Enter `ode45('heart',[0,2*pi],[0;2])` and note that MATLAB dynamically plots both y_1 and y_2 versus t.

   b) Enter `options=odeset('OutputFcn','odephas2')`, followed by `ode45('heart',[0,2*pi],[0;2],options)` and note that MATLAB plots y_2 versus y_1 in the phase plane (y_2 on the vertical axis, y_1 on the horizontal axis).

8. Use the technique of Example 3 to change the initial value problem $yy'' - (y')^2 - y^2 = 0$, $y(0) = 1$, $y'(0) = -1$ to a system with initial conditions. Use `ode45` to create a plot of y versus t.

9. An undriven spring-mass system with no damping can be modeled by the equation $y'' + ky = 0$, where k is the spring constant. However, suppose that the spring loses elasticity with time. A possible model for an "aging" spring-mass system is $y'' + 2e^{-0.12t}y = 0$. Suppose that the spring is stretched two units from equilibrium and released from rest. Use the technique of Example 3 to plot a solution of y versus t on the time interval [0, 100].

10. The equation $x'' = ax' - b(x')^3 - kx$ was devised by Lord Rayleigh to model the motion of the reed in a clarinet. With $a = 5$, $b = 4$, and $k = 5$, solve this equation numerically with initial conditions $x(0) = A$, and $x'(0) = 0$ over the interval [0, 10] for the three choices $A = 0.5$, 1, and 2. Use MATLAB's `hold` command to prepare both time plots and phase plane plots containing the solutions to all three initial value problems superimposed. Describe the relationship that you see between the three solutions.

11. Consider the driven, damped oscillator $y'' + 0.05y' + 36y = \sin 6.3t$, with initial conditions $y(0) = 5$, $y'(0) = 0$. Use the technique of Example 3 to transform the equation to system of first order ODEs, write and save the odefile as `osc.m`, and solve the system with the command `[t,x]=ode45('osc',[0,100],[5;0]);`.

    a) The command `plot3(x(:,1),x(:,2),t)` provides a three dimensional plot with time on the vertical axis.

    b) The command `close all, comet3(x(:,1),x(:,2),t)` provides an animated view of the result in part (a).

120

12. Consider the driven, undamped oscillator $y'' + \omega_0{}^2 y = 2\cos\omega t$. The parameter ω_0 is the *natural frequency* of the undriven system $y'' + \omega_0{}^2 y = 0$. If the natural frequency of the system is $\omega_0 = 2$, then the driven oscillator becomes $y'' + 4y = 2\cos\omega t$. Write the ODE file

```
function xprime=res(t,x,flag,omega)
xprime=zeros(2,1);
xprime(1)=x(2);
xprime(2)=-4*x(1)+2*cos(omega*t);
```

and save the file as `res.m`.

a) If the driving frequency nearly matches the natural frequency, then a phenomenon called *beats* occurs. Try `ode45('res',[0,60*pi],[0;0],[],1.9)`.

b) If the driving frequency matches the natural frequency, then a phenomenon called *resonance* occurs. Try `ode45('res',[0,60*pi],[0;0],[],2)`.

c) Use the `plot3` and `comet3` commands, as described in Exercise 11, to create three dimensional plots of the output in parts (a) and (b).

13. **Harmonic motion.**

An undriven, damped oscillator has equation

$$my'' + cy' + ky = 0,$$

where m is the mass, c is the damping constant, and k is the spring constant. Write the equation as a system of first order ODEs and create a function ODE file that allows the passing of parameters m, c, and k. In each of the following cases compute the solution with initial conditions $y(0) = 1$, and $y'(0) = 0$ over the interval $[0, 20]$. Prepare both a plot of y versus t and a phase plane plot of y' versus y.

a) (No damping) $m = 1, c = 0$, and $k = 16$.

b) (Under damping) $m = 1, c = 2$, and $k = 16$.

c) (Critical damping) $m = 1, c = 8$, and $k = 16$.

d) (Over damping) $m = 1, c = 10$, and $k = 16$.

14. The system

$$\varepsilon \frac{dx}{dt} = x(1-x) - \frac{(x-q)}{(q+x)}fz,$$

$$\frac{dz}{dt} = x - z,$$

models something chemists call an oregonator. Suppose $\varepsilon = 10^{-2}$ and $q = 9 \times 10^{-5}$. Write the ODE file

```
function xprime=oreg2(t,x,flag,f)
xprime=zeros(2,1);
xprime(1)=(x(1)*(1-x(1))-f*x(2)*(x(1)-9e-5)/(9e-5+x(1)))/1e-2;
xprime(2)=x(1)-x(2);
```

and save it as `oreg2.m`.

a) The command `ode45('oreg2',[0,50],[0.2;0.2],[],1/4)` provides no difficulties.

b) The idea is to vary the parameter f and note its affect on the oregonator model. However, `ode45('oreg2',[0,50],[0.2;0.2],[],1)` sets off all kinds of warning messages. Try it!

c) Try to improve the accuracy with `options=odeset('RelTol',1e-6)` and `ode45('oreg2',[0,50],[0.2;0.2],options,1)`, but this slows computation to a crawl as the system is very stiff. Try it!

d) The secret is to use `ode15s`. Then you can note the oscillations in the reaction. Try
`ode15s('oreg2',[0,50],[0.2;0.2],options,1)`.

15. Consider the oregonator model

$$\frac{dx}{dt} = \frac{qy - xy + x(1 - x)}{\varepsilon},$$

$$\frac{dy}{dt} = \frac{-qy - xy + fz}{\varepsilon'},$$

$$\frac{dz}{dt} = x - z.$$

Set $\varepsilon = 10^{-2}$, $\varepsilon' = 2.5 \times 10^{-5}$, and $q = 9 \times 10^{-5}$. Set up an ODE file that will allow the passing of the parameter f. Slowly increase the parameter f from a low of 0 to a high of 1 until you clearly see the system oscillating on the interval [0, 50] with initial conditions $x(0) = 0.2$, $y(0) = 0.2$, and $z(0) = 0.2$. You might want to read the comments in Exercise 14. Also, you'll want to experiment with `plot`, `zoom`, and `semilogy` to best determine how to highlight the oscillation in all three components.

16. The system

$$m_1 x'' = -k_1 x + k_2(y - x),$$

$$m_2 y'' = -k_2(y - x),$$

models a *coupled* oscillator. Imagine a spring (with spring constant k_1) attached to a hook in the ceiling. Mass m_1 is attached to the spring, and a second spring (with spring constant k_2) is attached to the bottom of mass m_1. If a second mass, m_2, is attached to the second spring, you have a coupled oscillator, where x and y represent the displacements of masses m_1 and m_2 from their respective equilibrium positions.

If you set $x_1 = x$, $x_2 = x'$, $x_3 = y$, and $x_4 = y'$, you can show that

$$x_1' = x_2,$$

$$x_2' = -\frac{k_1}{m_1} x_1 + \frac{k_2}{m_1}(x_3 - x_1),$$

$$x_3' = x_4,$$

$$x_4' = -\frac{k_2}{m_2}(x_3 - x_1).$$

Assume $k_1 = k_2 = 2$ and $m_1 = m_2 = 1$. Create an ODE file and name it `couple.m`. Suppose that the first mass is displaced upward two units, the second downward two units, and both masses are released from rest. Plot the position of each mass versus time.

17. In the 1920's, the Italian mathematician Umberto Volterra proposed the following mathematical model of a predator-prey situation to explain why, during the first World War, a larger percentage of the catch of Italian fishermen consisted of sharks and other fish eating fish than was true both before and after the war. Let $x(t)$ denote the population of the prey, and let $y(t)$ denote the population of the predators.

In the absence of the predators, the prey population would have a birth rate greater than its death rate, and consequently would grow according to the exponential model of population growth, i.e. the growth rate of the population would be proportional to the population itself. The presence of the predator population has the effect of reducing the growth rate, and this reduction depends on the number of encounters between individuals of the two species. Since it is reasonable to assume that the number of such encounters is proportional to the number of individuals of each population, the reduction in the growth rate is also proportional to the product of the two populations, i.e., there are constants a and b such that

$$x' = ax - bxy. \tag{7.22}$$

Since the predator population depends on the prey population for its food supply it is natural to assume that in the absence of the prey population, the predator population would actually decrease, i.e. the growth rate would be negative. Furthermore the (negative) growth rate is proportional to the population. The presence

122

of the prey population would provide a source of food, so it would increase the growth rate of the predator species. By the same reasoning used for the prey species, this increase would be proportional to the product of the two populations. Thus, there are constants c and d such that

$$y' = -cy + dxy. \tag{7.23}$$

a) A typical example would be with the constants given by $a = 0.4$, $b = 0.01$, $c = 0.3$, and $d = 0.005$. Start with initial conditions $x_1(0) = 50$ and $x_2(0) = 30$, and compute the solution to (7.22) and (7.23) over the interval [0, 100]. Prepare both a time plot and a phase plane plot.

After Volterra had obtained his model of the predator-prey populations, he improved it to include the effect of "fishing," or more generally of a removal of individuals of the two populations which does not discriminate between the two species. The effect would be a reduction in the growth rate for each of the populations by an amount which is proportional to the individual populations. Furthermore, if the removal is truly indiscriminate, the proportionality constant will be the same in each case. Thus, the model in equations (7.22) and (7.23) must be changed to

$$\begin{aligned} x' &= ax - bxy - ex, \\ y' &= -cy + dxy - ey, \end{aligned} \tag{7.24}$$

where e is another constant.

b) To see the effect of indiscriminate reduction, compute the solutions to the system in (7.24) when $e = 0$, 0.01, 0.02, 0.03, and 0.04, and the other constants are the same as they were in part a). Plot the five solutions on the same phase plane, and label them properly.

c) Can you use the plot you constructed in part b) to explain why the fishermen caught more sharks during World War I? You can assume that because of the war they did less fishing.

Student Projects

The following problems are quite involved, with multiple parts, and thus are good candidates for individual student projects. They also hold good potential for group projects, should students or instructors care to tackle these problems with student teams.

1. **The Non-linear Spring and Duffing's Equation.**

A more accurate description of the motion of a spring is given by *Duffing's equation*

$$my'' + cy' + ky + ly^3 = F(t).$$

Here m is the mass, c is the damping constant, k is the spring constant, and l is an additional constant which reflects the "strength" of the spring. Hard springs satisfy $l > 0$, and soft springs satisfy $l < 0$. As usual, $F(t)$ represents the external force.

Duffing's equation cannot be solved analytically, but we can obtain approximate solutions numerically in order to examine the effect of the additional term ly^3 on the solutions to the equation. For the following exercises, assume that $m = 1$ kg, that the spring constant $k = 16$ N/m, and that the damping constant is $c = 1$ kg/sec. The external force is assumed to be of the form $F(t) = A\cos(\omega t)$, measured in Newtons, where ω is the frequency of the driving force. The natural frequency of the spring is $\omega_0 = \sqrt{k/m} = 4$ rad/sec.

a) Let $l = 0$ and $A = 10$. Compute the solution with initial conditions $y(0) = 1$, and $y'(0) = 0$ on the interval [0, 20], with $\omega = 3.5$ rad/sec. Print out a graph of this solution. Notice that the steady state part of the solution dominates the transient part when t is large.

b) With $l = 0$ and $A = 10$, compute the amplitude of the steady state as follows. The amplitude is the maximum of the values of $|y(t)|$. Because we want the amplitude of the steady state oscillation, we only want to allow large values of t, say $t > 15$. This will allow the transient part of the solution to decay. To compute this number in MATLAB, do the following. Suppose that Y is the vector of y-values, and T is the vector of corresponding values of t. At the MATLAB prompt type

```
max(abs(Y.*(T>15)))
```

Your answer will be a good approximation to the amplitude of the steady state solution.

Why is this true? The expression (T>15) yields a vector the same size as T, and an element of this vector will be 1 if the corresponding element of T is larger than 15 and 0 otherwise. Thus Y.*(T>15) is a vector the same size as T or Y with all of the values corresponding to $t \le 15$ set to 0, and the other values of Y left unchanged. Hence, the maximum of the absolute values of this vector is just what we want.

Set up a script M-file that will allow you to do the above process repeatedly. For example, if the derivative M-file for Duffing's equation is called duff.m, the script M-file could be

```
[t,y]=ode45('duff',0,20,[1,0]);
y = y(:,1);
amplitude = max(abs(y.*(t>15)))
```

Now by changing the value of ω in duff.m you can quickly compute how the amplitude changes with ω. Do this for eight evenly spaced values of ω between 3 and 5. Use plot to make a graph of amplitude vs. frequency. For approximately what value of ω is the amplitude the largest?

c) Set $l = 1$ (the case of a hard spring) and $A = 10$ in Duffing's equation and repeat b). Find the value of ω, accurate to 0.2 rad/sec, for which the amplitude reaches its maximum.

d) For the hard spring in c), set $A = 40$. You are to redo c), but with two different choices of initial conditions, and for eight evenly spaced values between 5 and 7. The initial conditions are $y(0) = y'(0) = 0$, and $y(0) = 6$, $y'(0) = 0$. Plot the two graphs of amplitude vs. frequency on the same figure. (The phenomenon you will observe is called Duffing's hysteresis.)

e) Set $l = -1$ and $A = 10$ (the case of a soft spring) in Duffing's equation and repeat c).

2. **The Lorenz System.**

The purpose of this exercise is to explore the complexities displayed by the Lorenz system as the parameters are varied. We will keep $a = 10$, and $b = 8/3$, and vary r. For each value of r, examine the behavior of the solution with different initial conditions, and make conjectures about the limiting behavior of the solutions as $t \to \infty$. The solutions should be plotted in a variety of ways in order to get the information. These include time plots, such as Figure 7.5, and phase space plots, such as Figure 7.6. In the case of phase space plots, use the rotate3d command to capture a view of the system that provides the best visual information. You might also use other plots, such as z versus x, etc.

Examine the Lorenz system for a couple of values of r in each of the following intervals. Describe the limiting behavior of the solutions.

a) $0 < r < 1$.

b) $1 < r < 470/19$. There are two distinct cases.

c) $470/19 < r < 130$. The case done in the text. This is a region of chaotic behavior of the solutions.

d) $150 < r < 165$. Things settle down somewhat.

e) $215 < r < 280$. Things settle down even more.

3. **Motion Near the Lagrange Points.**

Consider two large spherical masses of mass $M_1 > M_2$. In the absence of other forces, these bodies will move in elliptical orbits about their common center of mass (if it helps, think of these as the earth and the moon). We will consider the motion of a third body (perhaps a spacecraft), with mass which is negligible in comparison to M_1 and M_2, under the gravitational influence of the two larger bodies. It turns out that there are five equilibrium points for the motion of the small body relative to the two larger bodies. Three of these were found by Euler, and are on the line connecting the two large bodies. The other two were found by Lagrange and are called the Lagrange points. Each of these forms an equilateral triangle in the plane of motion with the positions of the two large bodies. We are interested in the motion of the spacecraft when it starts near a Lagrange point.

In order to simplify the analysis, we will make some assumptions, and choose our units carefully. First we will assume that the two large bodies move in circles, and therefore maintain a constant distance from each other. We will take the origin of our coordinate system at the center of mass, and we will choose rotating coordinates, so that the x-axis always contains the two large bodies. Next we choose the distance between the large bodies to be the unit of distance, and the sum of the the two masses to be the unit of mass. Thus $M_1 + M_2 = 1$. Finally we choose the unit of time so that a complete orbit takes 2π units; i.e., in our units, a year is 2π units. This last choice is equivalent to taking the gravitational constant equal to 1.

With all of these choices, the fundamental parameter is the relative mass of the smaller of the two bodies

$$\mu = \frac{M_2}{M_1 + M_2} = M_2.$$

Then the location of M_1 is $(-\mu, 0)$, and the position of M_2 is $(1 - \mu, 0)$. The position of the Lagrange point is $((1 - 2\mu)/2, \sqrt{3}/2)$. If (x, y) is the position of the spacecraft, then the distances to M_1 and M_2 are

$$r_1^2 = (x + \mu)^2 + y^2,$$
$$r_2^2 = (x - 1 + \mu)^2 + y^2.$$

Finally, Newton's equations of motion in this moving frame are

$$x'' - 2y' - x = -(1 - \mu)(x + \mu)/r_1^3 - \mu(x - 1 + \mu)/r_2^3,$$
$$y'' + 2x' - y = -(1 - \mu)y/r_1^3 - \mu y/r_2^3. \tag{7.25}$$

a) Find a system of four first order equations which is equivalent to system (7.25).

b) If the two bodies are the earth and the moon, $\mu = 0.0122$. The Lagrange points are stable for $0 < \mu < 1/2 - \sqrt{69}/18 \approx 0.03852$, and in particular for the earth/moon system. Write $x = (1 - 2\mu)/2 + \xi$, and $y = \sqrt{3}/2 + \eta$, so (ξ, η) is the position of the spacecraft relative to the Lagrange point. Starting with initial conditions which are less than $1/100$ unit away from the Lagrange point, compute the solution. For each solution that you compute, make a plot of η vs. ξ to see the motion relative to the Lagrange point, and a plot of y vs. x, which also includes the positions of M_1 and M_2 to get a more global view of the motion.

c) Examine the range of stability by computing and plotting orbits for $\mu = 0.037$ and $\mu = 0.04$.

d) What is your opinion? Assuming μ is in the stable range, are the Lagrange points just stable, or are they asymptotically stable?

e) Find all five equilibrium points for the system you found in a). This is an algebraic problem of medium difficulty. It is not a computer problem unless you can figure out how to get the Symbolic Toolbox to find the answer.

f) Show that the equilibrium points on the x-axis are always unstable. This is a hard algebraic problem.

g) Show that the Lagrange points are stable for $0 < \mu < 1/2 - \sqrt{69}/18$. Decide whether or not these points are asymptotically stable. This is a very difficult algebraic problem.

8.   The Symbolic Toolbox

The Symbolic Toolbox provides a link between MATLAB and the symbolic algebra program known as Maple. Unlike the rest of MATLAB, which provides powerful numerical routines, the Symbolic Toolbox deals with symbols, formulas, and equations. In dealing with differential equations, it leads to explicit formulas for the solutions, provided such formulas exist at all.

The Symbolic Toolbox is included in Version 5 of the Student Edition of MATLAB, however, it is distributed as an additional toolbox in the standard version of MATLAB. Therefore it may not be available on your platform. If you are unsure whether or not it is available, execute the command `help symbolic` at the MATLAB prompt. If the Symbolic Toolbox is available, you will get a list of all of the commands available in the toolbox. If it is not, you will get an error message, saying `symbolic not found`.

For a general introduction to the Symbolic Toolbox, find a copy of the *Symbolic Math Toolbox User's Guide*. This provides a very good tutorial on the use of the Symbolic Toolbox. In some ways it provides a much more comprehensive treatment than we will present here.

In this chapter, we will put our emphasis on using the Symbolic Toolbox to solve differential equations. This means that MATLAB will try to tell you the exact, analytic solution to an equation, or even to an initial value problem. MATLAB can be successful, of course, only if such a solution exists.

Symbolic Objects

With the release of MATLAB 5, users now have the ability to add new data types to MATLAB by creating *classes*[1]. The class of a variable describes the structure of the variable and indicates the kinds of operations and functions that can apply to the variable. An *object* is an *instance* of a particular class.

The Symbolic Toolbox defines a `sym` class and you can create instances (objects) of this class with the `sym` command. For example, the commands

```
>> x=5;y=sym('alpha');
```

create two objects: `x` is an array object of class `double` with value 5, and `y` is an object of class `sym` with value `alpha`. The `whos` command reveals the class of objects in your workspace.

```
>> whos
  Name      Size      Bytes  Class

  x         1x1           8  double array
  y         1x1         134  sym object
```

The last column of the output from the `whos` command lists the class of objects in your workspace. You can examine the contents of objects in your workspace in the usual manner.

---

[1] For a detailed account of object-orientated programming, see Chapter 14, *Classes and Objects*, in *Using Matlab—Version 5*, the user's guide that accompanies the software distribution.

```
>> x
x =
     5
>> y
y =
alpha
```

There is a subtle distinction between the output of an object of class double and the output of an object of class sym: the sym object's contents are left justified, while the object of class double indents its contents about half an inch from the left edge of the command window.

In the previous example, the sym command was used to create an instance of class sym. However, you can also use the sym command to change the class of an existing object. For example, the command

```
>> A=hilb(3)
A =
     1.0000    0.5000    0.3333
     0.5000    0.3333    0.2500
     0.3333    0.2500    0.2000
```

creates an object A of class double, which is readily checked with the command

```
>> whos
  Name      Size          Bytes  Class

  A         3x3              72  double array
  x         1x1               8  double array
  y         1x1             126  sym object
```

You can change the class of this object to sym with the command

```
>> A=sym(A)
A =
[   1, 1/2, 1/3]
[ 1/2, 1/3, 1/4]
[ 1/3, 1/4, 1/5]
```

Check the class of A by typing whos at the MATLAB prompt[2]. Note the left justification of the sym object A.

You can add new classes to your MATLAB environment by specifying a MATLAB structure that provides data storage for the object and creating a directory of M-files that operate on the object. These M-files are known as the *methods* for the class. You can obtain a list of methods for objects of type sym by typing help symbolic at the MATLAB prompt.

Furthermore, you can redefine how a built-in MATLAB operator works for your class. This technique is known as *overloading* the operator. For example, the diff command will operate in a manner determined by the class of its arguments. If the input is a vector of class double, MATLAB takes differences of consecutive elements of the vector.

---

[2] You can also enter class(A) at the MATLAB prompt to identify the class of A. Try it!

```
>> x=[1 2 4 7 11]
x =
     1     2     4     7    11
>> diff(x)
ans =
     1     2     3     4
```

The creators of the Symbolic Toolbox have overloaded the `diff` operator; they have redefined how `diff` works for objects of class `sym`.

```
>> x=sym('x');
>>diff(x)
ans =
1
```

In this case, MATLAB recognizes that the input argument has class `sym` and promptly takes the derivative of x with respect to x, which is 1.

The introduction of object-orientated programming has necessitated a change in how the help is organized for MATLAB's operators and methods. Type `help diff` at the MATLAB prompt. If you read the help file carefully, you will note that there is no mention of symbolic differentiation. However, at the end of the file, you will note an announcement that the `diff` operator has been overloaded for objects of class `sym` and `char`. Type `help sym/diff` at the MATLAB prompt. Note that the resulting help file carefully discusses symbolic differentiation.

The Default Symbolic Variable

You can declare symbolic objects one at a time, such as

```
>> x=sym('x');
>> a=sym('a');
>> b=sym('b');
>> c=sym('c');
```

or you can use the `syms` command to declare them all at once.

```
>> syms x a b c
```

The order of declaration is unimportant.

The creators of the Symbolic Toolbox have overloaded the operators +, -, *, /, and ^, so you can use these operators to create the expressions needed in a standard differential equations course.

```
>> f=a^2*x^2+b*x+c;
```

Because a, b, c, and x have been declared to be symbolic objects, the methods of the Symbolic Toolbox will now recognize that f is a symbolic object and act accordingly.

```
>> diff(f)
ans =
2*a^2*x+b
```

128

You can just as easily take a second derivative.

```
>> diff(f,2)
ans =
2*a^2
```

How did the Symbolic Toolbox know to differentiate with respect to x, and not with respect to another symbolic variable, say a? The answer lies in the Symbolic Toolbox's *default variable*.

The findsym Rule. *The default symbolic variable in a symbolic expression is the letter that is closest to 'x' alphabetically. If there are two equally close, the letter later in the alphabet is chosen.*

The findsym command will find all of the symbolic variables in a symbolic expression.

```
>> findsym(f)
ans =
a, b, c, x
```

The command findsym(f,n) returns the n symbolic variables 'closest' to x. Hence,

```
>> findsym(f,1)
ans =
x
```

returns the default symbolic variable. That is why the Symbolic Toolbox differentiated with respect to x in the above example. Of course, most Symbolic Toolbox commands allow the user to override the default symbolic variable. For example,

```
>> diff(f,a)
ans =
2*a*x^2
```

finds the derivative of f with respect to a, and

```
>> diff(f,a,2)
ans =
2*x^2
```

finds the second derivative of f with respect to a.

Verifying Solutions to Differential Equations

After having completed extensive paper and pencil work to obtain the solution of a differential equation or initial value problem, you might want to check your answer. Why not let the Symbolic Toolbox do the work for you?

Example 1. *Verify that $y(t) = Ce^{-t^2}$ is a solution of*

$$y' + 2ty = 0, \tag{8.1}$$

where C is an arbitrary constant.

129

Declare the symbolic variables and enter the potential solution.

```
>> syms y t C
>> y=C*exp(-t^2);
```

It is simple to check that you have a true solution.

```
>> diff(y,t)+2*t*y
ans =
0
```

Example 2. *Verify that* $y = x \tan x$ *is a solution of*

$$xy' = y + x^2 + y^2. \tag{8.2}$$

Note that the independent variable is x, then declare symbolic variables and enter the potential solution.

```
>> syms x y
>> y=x*tan(x);
```

You can enter each side of equation (8.2) and compare, or you can enter $xy' - (y + x^2 + y^2)$ and verify that it equals zero.

```
>> x*diff(y,x)-(y+x^2+y^2)
ans =
x*(tan(x)+x*(1+tan(x)^2))-x*tan(x)-x^2-x^2*tan(x)^2
```

We didn't get zero! Or did we? Recall that `ans` always contains the answer to the last computation and use the Symbolic Toolbox's `simplify` command.

```
>> simplify(ans)
ans =
0
```

Therefore, $y = x \tan x$ is a solution of equation (8.2).

Solving Ordinary Differential Equations

The routine `dsolve` is probably the most useful differential equations tool in the Symbolic Toolbox. To get a quick description of this routine, enter `help dsolve`. Perhaps the best way to explain the syntax of the `dsolve` command is to start is with an example.

Example 3. *Find the general solution of the first order differential equation*

$$\frac{dy}{dt} + y = te^t. \tag{8.3}$$

The Symbolic Toolbox easily finds the general solution of the ODE (8.3)

```
>> dsolve('Dy+y=t*exp(t)')
ans =
1/4*(2*exp(2*t)*t-exp(2*t)+4*C1)/exp(t)
```

The routine `dsolve` requires that the differential equation be entered as a string, delimited with single apostrophes. The notation Dy is used for y', D2y for y'', etc.

The `dsolve` routine assumes that the independent variable is t. You can override this default by placing the independent variable of choice as the last input argument to `dsolve`. For example, `dsolve('Dy+y=x*exp(x)','x')` declares that the independent variable is x. Try it! It is probably a good practice to always declare the independent variable when using the `dsolve` command.

Note that the output was not assigned to the variable y, as one might expect. This is easily rectified.

```
>> y=dsolve('Dy+y=t*exp(t)','t')
y =
1/4*(2*exp(2*t)*t-exp(2*t)+4*C1)/exp(t)
```

The commands

```
>> y=expand(y)
y =
1/2*t*exp(t)-1/4*exp(t)+1/exp(t)*C1
```

```
>> pretty(y)

                                            C1
                   1/2 t exp(t) - 1/4 exp(t) + ------
                                            exp(t)
```

give a nicer display of the output.

Let's substitute a number for the constant C1 and obtain a plot of the result on the time interval $[-1, 3]$. The commands

```
>> C1=1
C1 =
     1
>> y=subs(y)
y =
1/2*t*exp(t)-1/4*exp(t)+1/exp(t)
```

replace the variable C1 in the symbolic expression y with the value of C1 in MATLAB's workspace[3]. MATLAB's `ezplot` command will produce an image similar to that in Figure 8.1.

---

[3] MATLAB's `subs` command is a multi-faceted, complex command and we will revisit it many times in this chapter. If any of the symbolic variables of the expression contained in the variable y have values in MATLAB's workspace, then the command `y=subs(y)` substitutes each of these values into the expression y, and assigns the resulting expression to the variable y.

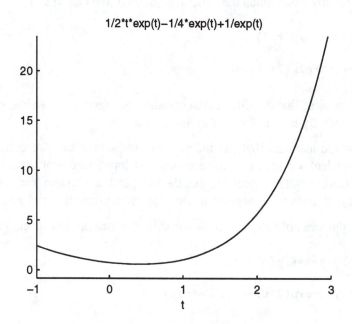

Figure 8.1. The plot of $(1/2)te^t - (1/4)e^t + e^{-t}$ on $[-1, 3]$.

```
>> ezplot(y,[-1,3])
```

Example 4. *Consider the equation of an undamped oscillator with periodic forcing function*

$$y'' + 144y = \cos 11t, \tag{8.4}$$

with initial position given by $y(0) = 0$ and initial velocity given by $y'(0) = 0$. Sketch a solution of this oscillator with these given initial conditions.

The Symbolic Toolbox easily finds solutions of initial value problems.

```
>> y=dsolve('D2y+144*y=cos(11*t)','y(0)=0,Dy(0)=0','t');
```

If you remove the semicolon suppressing the output of this last command, you will experience first hand the sometimes horribly complicated output delivered by computer algebra systems. If you type `simple(y)` you can watch as MATLAB attempts to simplify this output. The `simple` command executes a number of simplification techniques (count them) and somewhat arbitrarily decides that the best form of an expression is the one that is shortest in length. Of course, the shortest expression is not always the best simplification, but this strategy works surprisingly well.

If you type

```
>> y=simple(y)
y =
1/23*cos(11*t)-1/23*cos(12*t)
```

132

then the Symbolic Toolbox will suppress the various attempted simplifications, and respond only with the optimal simplification (again, the one shortest in length). That simplification is then assigned to the variable y.

We can plot this solution on the time interval $[0, 6\pi]$ with the command

```
>> ezplot(y,[0,6*pi])
```

producing an image similar to that in Figure 8.2. The phenomenon displayed in Figure 8.2 is called *beats*, a condition present in undamped, forced oscillators that are nearly resonant.

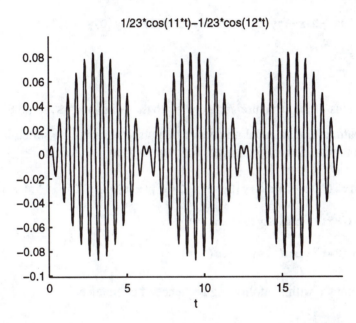

Figure 8.2. The solution of $y'' + 144y = \cos 11t$, $y(0) = 0$, $y'(0) = 0$.

Solving Systems of Ordinary Differential Equations

The routine `dsolve` easily handles systems of equations, but you might find `ezplot` lacking when you want to look at a visual display of the solution.

Example 5. *Find the solution of the initial value problem*

$$x' = -2x - 3y,$$
$$y' = 3x - 2y,$$

(8.5)

with $x(0) = 1$ *and* $y(0) = -1$.

It is a good idea to clear variables from your workspace when they are no longer needed. Left-over variables can wreak unexpected havoc when employing the subs command. You can delete all[4] variables from your workspace with

```
>> clear all
```

Enter the whos command to verify that there are no variables left in the workspace.

We can call dsolve in the usual manner. Readers will note that the input to dsolve in this example parallels that in Examples 3 and 4. The difference lies in the output variable(s). If a single variable is used to gather the output, then the Symbolic Toolbox returns a MATLAB structure[5] variable whose fields contain the solutions of system (8.5).

```
>> S=dsolve('Dx=-2*x-3*y,Dy=3*x-2*y','x(0)=1,y(0)=-1','t')
S =
    x: [1x1 sym]
    y: [1x1 sym]
```

You can access the fields of the structure variable S with the commands S.x and S.y. Try it!

For several equations and an equal number of outputs, the results are sorted in lexicographic order and assigned to the outputs. You might find the following syntax preferable.

```
>> [x,y]=dsolve('Dx=-2*x-3*y,Dy=3*x-2*y','x(0)=1,y(0)=-1','t')
x =
exp(-2*t)*cos(3*t)+exp(-2*t)*sin(3*t)
y =
exp(-2*t)*sin(3*t)-exp(-2*t)*cos(3*t)
```

Plot each solution versus t with the following sequence of commands.

```
>> ezplot(x,[0,2*pi])
>> hold on
>> ezplot(y,[0,2*pi])
```

Adjust the view with the axis command, and add a title and axis labels.

```
>> axis([0,2*pi,-1.2,1.2])
>> title('Solutions of x''=-2x-3y, y''=3x-2y, x(0)=1, y(0)=-1.')
>> xlabel('Time t')
>> ylabel('x and y')
```

This should produce an image similar to that in Figure 8.3.

---

[4] You can also select individual variables for deletion. For example, the command clear t clears only the variable t from your workspace.

[5] For a complete introduction to structures and cell arrays, see Chapter 13 in *Using Matlab — Version 5*, the user guide that accompanies the software.

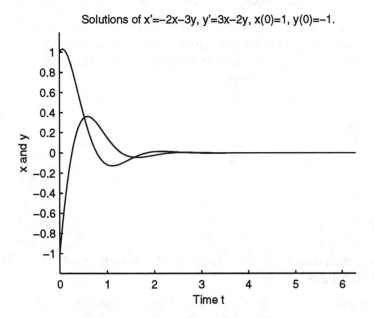

Figure 8.3. Plots of x and y versus t.

Handle Graphics. Unfortunately, the Symbolic Toolbox's `ezplot` command does not have the formatting capabilities of MATLAB's `plot` command. There is no built-in structure for changing the linestyle or the color of the plot. If you wish to format the plot in Figure 8.3, you have to use MATLAB's *handle graphics*.

Normally, we would refer you to MATLAB's documentation for a complete treatment of handle graphics. That's still good advice, but you don't have to fully understand everything about handle graphics techniques to effectively use them in spot situations. For example, use your mouse to click on one of the solution trajectories in Figure 8.3, then immediately type h=gco at the MATLAB prompt. The command `gco`, which means "get current object," stores a numerical handle to the last graphics object selected with the mouse. You can get a full set of properties (and a list of possible values for each) of the graphics object by typing `set(h)`. You can get current settings of the object with handle h with the command `get(h)`. Try each of these commands now!

Let's try to change the LineStyle and Color properties of the solution trajectory with handle h. The command `set(h,'Color','r')` should change the color of the solution trajectory to red. Try it! The command `set(h,'LineStyle','--')` should change the line style of the solution trajectory to dashed. Try it!

Experiment further with the `Marker`, `LineWidth`, and `MarkerSize` properties. One of the nicest ways to obtain a summary of these properties is to type `helpdesk`, then click the Handle Graphics Properties link when the MATLAB Help Desk opens in your browser. Click the link to Line objects and you can explore further some of the properties we mentioned above.

Phase Plot. However, if you want a plot in the phase plane, then you are out of luck. The Symbolic Toolbox has no immediate plotting routines that will do this for you. We need to use MATLAB's `plot` command, but that is a numerical routine. How are we going to generate numerical data for the `plot` command when both x and y are sym objects? The answer lies in the Symbolic Toolbox's subs command.

In Example 3, the command y=subs(y) substituted the value C1=1 in MATLAB's workspace into the expression contained in the variable y and assigned the resulting expression to the variable y. Although this is a powerful use of the subs command, sometimes this form does not provide enough control over the substitution process. For example, the code

```
>> syms x a b
>> x=a*x+b
x =
a*x+b
>> b=1;
>> subs(x)
ans =
a*(a*x+b)+1
```

provides more substitution than wanted (we just wanted to substitute 1 for b). What is needed is finer control over what gets substituted where.

For a complete description of the Symbolic Toolbox's subs command, type help subs at the MATLAB prompt and read the resulting helpfile. The syntax that we need to create our phase plane plot is subs(S,OLD,NEW), where

- S contains the symbolic expression to be manipulated,

- OLD is a symbolic variable, a string representing a variable name, or a string (quoted) expression, and

- NEW is a symbolic or numerical variable or expression.

A new symbolic expression is formed by substituting the contents of the variable NEW into each occurrence of the variable OLD in the symbolic expression S.

The subs command has been carefully crafted by the creators of the Symbolic Toolbox and will automatically make our solutions of system (8.5) array smart. Create a vector of time values you want to substitute into the expressions contained in the variables x and y.

```
>> td=linspace(0,2*pi);
```

One might try the following first attempt at substitution,

```
>> xd=subs(x,t,td)
??? Undefined function or variable 't'.
```

but MATLAB responds with an error message indicating that is has no knowledge of the variable t. A quick glance at the second bulleted item above reveals that the input argument OLD must be a symbolic variable or a string representing a variable name. The command xd=subs(x,'t',td) will work, but so will

```
>> syms t
>> xd=subs(x,t,td);
>> yd=subs(y,t,td);
```

136

It is illuminating to unsuppress the output (remove the semi-colons) in the above commands. A quick check of MATLAB's workspace with the whos command will also reveal that xd and yd are arrays of class double; that is, arrays of double precision floating point numbers.

The commands

```
>> subplot(211)
>> plot(td,xd,td,yd)
>> xlabel('t')
>> ylabel('x and y')
>> subplot(212)
>> plot(xd,yd)
>> xlabel('x')
>> ylabel('y')
```

will produce the x and y versus t plots, and the phase plane plot of y versus x shown in Figure 8.4.

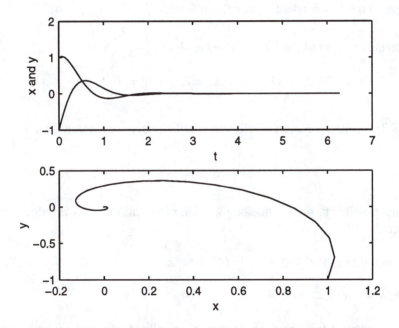

Figure 8.4. Plots of x and y versus t and a phase plane plot.

For an animated plot, try close all, comet(xd,yd).

Interpreting the Output of DSOLVE

Dsolve can output some strange looking solutions in certain situations.

Example 6. *Find the solution of the differential equation*

$$2tyy' = 3y^2 - t^2. \tag{8.6}$$

Note that equation (8.6) is nonlinear. Exact solutions of nonlinear differential equations are rare, but the Symbolic Toolbox will find them if they exist.

```
>> y=dsolve('2*t*y*Dy=3*y^2-t^2','t')
y =
[ -(1+t*C1)^(1/2)*t]
[  (1+t*C1)^(1/2)*t]
```

The output y is a vector. You can access each component of this vector in the usual manner; that is, y(1) is the first component, and y(2) is the second component. Try it!

This answer seems unusual until we find the solution ourselves. The substitution $y = tz$ yields a separable equation whose solution is $z^2 - 1 = Ct$, where C is an arbitrary constant. Setting $z = y/t$ yields the implicit solution $y^2 = Ct^3 + t^2$, and solving this equation explicitly for y yields the solutions $y = \pm t\sqrt{1 + Ct}$. These are the solutions stored in the vector y.

Sometimes the Symbolic Toolbox cannot find explicit (where the independent variable is expressed explicitly in terms of the independent variable) solutions.

Example 7. *Find the solution of the initial value problem*

$$(1 - \sin x)x' = t, \quad x(0) = 0. \tag{8.7}$$

Equation (8.7) is separable and the following implicit solution is easily found.

$$x + \cos x = \frac{t^2}{2} + 1 \tag{8.8}$$

Of course, it is not possible to solve equation (8.8) for x in terms of t, so the output from dsolve is not entirely unexpected.

```
>> x=dsolve('(1-sin(x))*Dx=t','x(0)=0','t')
x =
RootOf(2*_Z+2*cos(_Z)-t^2-2)
```

This rather cryptic output is telling us that an explicit solution could not be found. Further, the notation RootOf(2*_Z+2*cos(_Z)-t^2-2) informs us that the implicit solution is a root of the equation $2Z + 2\cos Z - t^2 - 2 = 0$. If you multiply both sides of equation (8.8) by 2 and move all terms to the left-hand side, you get $2x + 2\cos x - t^2 - 2 = 0$, making the solution x a root of the equation $2Z + 2\cos Z - t^2 - 2 = 0$.

Consequently, the notation RootOf(2*_Z+2*cos(_Z)-t^2-2) represents the implicit solution $2x + 2\cos x - t^2 - 2 = 0$.

The Solve Command

The solve command is the Symbolic Toolbox's versatile equation solver. It performs equally well with single equations and multiple equations. For example, if you need to solve the equation $ax + b = 0$, then the command

```
>> solve('a*x+b')
ans =
-b/a
```

provides the solution, $x = -b/a$. We need to make two important points about the syntax of the `solve` command. First, the command `solve('a*x+b')` is identical to the command `solve('a*x+b=0')`. Try it! Second, we did not indicate what variable to solve for, so the `solve` command found a solution for the default variable[6]. You can choose a different solution variable. The command

```
>> solve('a*x+b','a')
ans =
-b/x
```

solves the equation $ax + b = 0$ for a.

If an equation has multiple solutions, they are listed in an output vector.

```
>> s=solve('a*x^2+b*x+c')
s =
[ 1/2/a*(-b+(b^2-4*a*c)^(1/2))]
[ 1/2/a*(-b-(b^2-4*a*c)^(1/2))]
```

You can access the components of the solution vector in the usual manner.

```
>> pretty(s(1))
                         2           1/2
                   -b + (b  - 4 a c)
             1/2 --------------------
                          a
```

Finally, if the Symbolic Toolbox cannot find a symbolic solution, it will attempt to find a numerical solution.

```
>> s=solve('x+2=exp(x)')
s =
[     -lambertw(-exp(-2))-2]
[ -lambertw(-1,-exp(-2))-2]
```

This somewhat cryptic response can be made more intuitive by changing the class of s.

```
>> double(s)
ans =
   -1.8414
    1.1462
```

You can ask for more significant digits with the Symbolic Toolbox's variable precision arithmetic.

```
>> vpa(s,10)
ans =
[ -1.841405660]
[  1.146193221]
```

---

[6] Type `findsym(sym('a*x+b'),1)` to find the default variable in the expression $ax + b$.

Example 8. *The logistic model for population growth is given by the equation*

$$\frac{dp}{dt} = k\left(1 - \frac{P}{N}\right)P, \tag{8.9}$$

where k is the natural growth rate of the population and N is the carrying capacity of the environment. Assume that k = 0.025 and N = 1000. How long will it take an initial population of 200 to grow to 800? Assume that time is measured in minutes.

With $k = 0.025$ and $N = 1000$, equation (8.9) becomes $dP/dt = 0.025(1 - P/1000)P$, with initial population given by $P(0) = 200$. Use the `dsolve` command to find a solution.

```
>> P=dsolve('DP=0.025*(1-P/1000)*P','P(0)=200','t')
P =
1000/(1+4*exp(-1/40*t))
```

Consequently, the solution of the initial value problem is

$$P = \frac{1000}{1 + 4e^{-t/40}}. \tag{8.10}$$

The following commands provide a plot similar to that in Figure 8.5.

```
>> ezplot(P,[0,300])
>> grid
>> xlabel('Time in minutes')
>> ylabel('Population')
>> title('A Logistic Model')
```

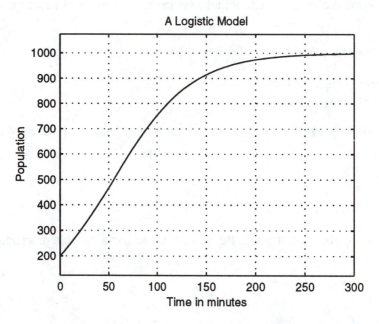

Figure 8.5. The solution of $dP/dt = 0.025(1 - P/1000)P$, $P(0) = 200$.

If you examine the plot in figure 8.5, it appears that the population reaches a level of 800 somewhere between 100 and 150 minutes.

If we substitute $P = 800$ in equation (8.10)

$$800 = \frac{1000}{1 + 4e^{-t/40}},$$

we can use the `solve` command[7] to find a more accurate solution.

```
>> s=solve('800=1000/(1+4*exp(-t/40))','t')
s =
40*log(16)
>> double(s)
ans =
   110.9035
```

Consequently, it takes approximately 111 minutes for the population to reach 800.

Exercises

1. In each of the following cases, use MATLAB and the technique demonstrated in Examples 1 and 2 to verify that y is a solution to the indicated equation.

   a) $y = 1 + e^{-t^2/2}$, $y' + ty = t$.

   b) $w = 1/(s - 3)$, $w' + w^2 = 0$.

   c) $y = 10 - t^2/2$, $yy' + ty = 0$.

   d) $y = \ln(x)$, $(2e^y - x)y' = 1$. Recall: $\ln(x)$ is `log(x)` in MATLAB.

   e) $y = e^t \cos(2t)$, $y'' - 2y' + 5y = 0$.

2. First determine the independent variable, then use `dsolve` to find the general solution to the following equations. Use the `subs` command to replace the integration constant C1 with C1=2. Use `ezplot` to plot the resulting solution.

   a) $y' + ty = t$.

   b) $y' + y^2 = 0$.

   c) $yy' + ty = 0$.

   d) $(2e^y - x)y' = 1$. Hint: Use `dsolve('(2*exp(y)-x)*Dy=1','x')`.

   e) $(x + y^2)y' = y$.

   f) $x(y' - y) = e^x$.

3. First determine the independent variable, then use `dsolve` to find the solution to the following initial value problems. Use `ezplot` to plot each solution over the indicated time interval.

   a) $y' + ty = t$, $y(0) = -1$, $[-4, 4]$.

   b) $y' + y^2 = 0$, $y(0) = 2$, $[0, 5]$.

   c) $yy' + ty = 0$, $y(1) = 4$, $[-4, 4]$.

   d) $(2e^y - x)y' = 1$, $y(0) = 0$, $[-5, 5]$. Hint: `dsolve('(2*exp(y)-x)*Dy=1','y(0)=0','x')`.

---

[7] You can actually get pretty fancy here and enter `solve(P-800)`. This is pretty sophisticated use of the Symbolic Toolbox, and indicates some of the power you can attain with experience.

e) $(x + y^2)y' = y$, $y(0) = 4$, $[-4, 6]$.

f) $x(y' - y) = e^x$, $y(1) = 4e$, $[0.0001, 0.001]$, and $[0.001, 1]$.

4. Use `dsolve` to obtain the solution of each of the following second order differential equations and the `simple` command to find the simplest form of that solution. Use `ezplot` to sketch the solution on the indicated time interval.

a) $y'' + 4y = 3\cos 2.1t$
$y(0) = 0$, $y'(0) = 0$, $[0, 64\pi]$

b) $y'' + 16y = 3\sin 4t$
$y(0) = 0$, $y'(0) = 0$, $[0, 32\pi]$

5. Use `dsolve` to obtain the solution of $y'' + y' + 144y = \cos 2t$, $y(0) = 0$, $y'(0) = 0$. Use `ezplot` to plot the solution on the time interval $[0, 6\pi]$. Use `ode45` to find a numerical solution on the same time interval and superimpose that plot of the first plot. How closely does the numerical solution match the symbolic solution?

6. Use the technique of Examples 1 and 2 to verify that

$$x = e^{-t}(\cos 2t - \sin 2t),$$
$$y = e^{-t}(3\sin 2t - \cos 2t),$$

is a solution of the system

$$x' = -5x - 2y,$$
$$y' = 10x + 3y.$$

Furthermore, use the `subs` command to verify that the solution satisfies the initial conditions $x(0) = 1$ and $y(0) = -1$.

7. Use `dsolve` to solve the initial value problem

$$y' = v,$$
$$v' = -2v - 2y + \cos 2t,$$

over the interval $[0, 3]$. Explain any apparent discontinuities you observe.

8. Find the solutions to the following initial value problems:

a) $t^2 y'' = (y')^2$, with $y(1) = 3$ and $y'(1) = 2$.

b) $y'' = yy'$, with $y(0) = 0$, and $y'(0) = 2$.

c) $y'' = ty' + y + 1$, with $y(0) = 1$, and $y'(0) = 0$.

9. Suppose we start with a population of 100 individuals at time $t = 0$, and that the population is correctly modeled by the logistic equation. Suppose that at time $t = 2$ there are 200 individuals in the population, and that the population reaches steady state with a population of 1000. Plot the population over the interval $[0, 20]$. What is the population at time $t = 10$?

10. Suppose we have a population which is correctly modeled by the logistic equation, and experimental measurements show that $p(0) = 50$, $p(1) = 150$, and $p(2) = 250$. Use the Symbolic Toolbox to derive the formula for the population as a function of time. Plot the population over the interval $[0, 5]$. What is the limiting value of the population? What is $p(3)$?

142

9. Linear Algebra Using MATLAB

MATLAB is short for "Matrix Laboratory." It contains a large number of built-in routines that make matrix algebra very easy. In this chapter we will learn how to use some of those routines to solve systems of linear equations.

Systems of Linear Equations

We are interested in solving systems of equations like

$$u + 2v = 5,$$
$$4u - v = 2. \tag{9.1}$$

This is a system of two equations involving two unknowns. We want to find a systematic method that will work in general for m equations involving n unknowns, where m, and n are possibly very large, and not necessarily equal.

The general method is not significantly different from the method usually taught in secondary school to solve systems like (9.1). We solve the first equation for the variable u, yielding $u = 5 - 2v$, and then we substitute this result into the second equation. In this way we obtain a new system consisting of the first equation and the modified second equation

$$u + 2v = 5,$$
$$-9v = -18, \tag{9.2}$$

which has the same solutions as (9.1). The advantage is that system (9.2) is quite easily solved. From the second equation, we see that $v = 2$. Then substituting $v = 2$ into the first equation, we can solve for u, getting $u = 1$.

Solving and substituting gets the job done, but this is difficult to implement for larger systems. The key to developing a more systematic approach is to notice that we can get the system in (9.2) by subtracting 4 times the first equation in (9.1) from the second equation in (9.1). This operation of adding (or subtracting) a multiple of one equation to (or from) another is equivalent to solving and substituting, but is more direct and easier to implement.

While the operation of adding a multiple of one equation of a system to another has been illustrated only for the special case of (9.1), it is easily seen that such an operation on an arbitrary system leads to a new system which is equivalent to the first in the sense that any solution of one system is also a solution of the other.

Our next step in devising a general solution method is to write system (9.1) as a matrix equation (See Example 3, Chapter 3).

$$\begin{bmatrix} 1 & 2 \\ 4 & -1 \end{bmatrix} \begin{bmatrix} u \\ v \end{bmatrix} = \begin{bmatrix} 5 \\ 2 \end{bmatrix} \tag{9.3}$$

The matrix equation (9.3) has the form $A\mathbf{x} = \mathbf{b}$, where

$$A = \begin{bmatrix} 1 & 2 \\ 4 & -1 \end{bmatrix}, \quad \mathbf{x} = \begin{bmatrix} u \\ v \end{bmatrix}, \quad \text{and} \quad \mathbf{b} = \begin{bmatrix} 5 \\ 2 \end{bmatrix}.$$

The matrix equation $A\mathbf{x} = \mathbf{b}$ is equivalent to the system of equations in (9.1). All of the information about the system is contained in the *coefficient matrix A*, and the *right hand side* vector $\mathbf{b}$. From these two we form the *augmented matrix M* by adding the column vector $\mathbf{b}$ as a third column to the matrix A; i.e.,

$$M = [A, \mathbf{b}] = \begin{bmatrix} 1 & 2 & 5 \\ 4 & -1 & 2 \end{bmatrix}.$$

The augmented matrix contains all of the information about the system in a compact form.

Notice that each row in M corresponds to one of the equations in (9.1). For example, the first row of M, [1 2 5], represents the equation $u + 2v = 5$. Furthermore, if we add -4 times the first row of M to the second row, we get

$$\begin{bmatrix} 1 & 2 & 5 \\ 0 & -9 & -18 \end{bmatrix},$$

which is the augmented matrix corresponding to the system in (9.2). Thus, we see that the method of solving and substitution in the system of equations becomes the operation of adding a multiple of one row of the augmented matrix to another row.

This is our first example of a *row operation* on a matrix. There are three in all.

Theorem 1. *Each of the following elementary row operations, when applied to the augmented matrix of a system of equations, will transform the augmented matrix into the augmented matrix of another system of equations which has exactly the same solution as the original system.*

R1. *Add a multiple of one row in a matrix to another row.*
R2. *Interchange two rows of the matrix.*
R3. *Multiply a row of a matrix by a nonzero constant.*

If the matrix is the augmented matrix associated to a system of equations, then operation R2 corresponds to interchanging two of the equations, and R3 corresponds to multiplying an equation by a nonzero constant.

Our general strategy for solving systems is to replace the system by the augmented matrix, and then perform row operations until the system associated with the transformed matrix is easy to solve. Of course, each of the row operations requires a significant amount of arithmetic, but we will let MATLAB do that arithmetic for us. We will discuss how to do this in MATLAB after addressing some preliminaries.

Matrix Indexing in MATLAB and Row Operations

One of the key features of MATLAB is the ease with which the elements of a matrix can be accessed and manipulated. We will explain some of these features here, in preparation for our explanation of how to implement row operations on matrices.

If M is a matrix in MATLAB, then the entry in the second row and fourth column is denoted by M(2,4). The basic idea is that the first index refers to the row, and the second to the column. Try the following example.

```
>> M=[1 3 5 6;3 5 0 -3;-4 0 9 3]
M =
     1     3     5     6
     3     5     0    -3
    -4     0     9     3

>> M(2,4)
ans =
    -3
```

Try $M(1,3)$. How would you access the entry in row three, column four, of matrix M?

The entries in a matrix can be easily changed. For example, if we want to change $M(3,4)$ to -15, we enter

```
>> M(3,4)=-15
M =
     1     3     5     6
     3     5     0    -3
    -4     0     9   -15
```

and the job is done.

We can refer to the second row in the matrix M by $M(2,:)$. The colon means we want all columns of row two, or in other words, all of the second row.

```
>> M(2,:)
ans =

     3     5     0    -3
```

The command $>> M(:,2)$ displays the second column of M. Try it! The colon indicates that we want all rows of column two.

We can also easily refer to submatrices. For example, if we want to refer to the matrix consisting of the first and third rows of M, we enter

```
>> M([1 3],:)
ans =

     1     3     5     6
    -4     0     9   -15
```

and we get what we want. Of course, we can do the same with columns.

```
>> M(:,[2,4])
ans =

     3     6
     5    -3
     0   -15
```

What do you think M([1,3],[4,1]) refers to? Try it and see.

Now we are ready to explain how to do row operations. We will start with the matrix M we ended up with above, i.e.

```
>> M
M =

     1     3     5     6
     3     5     0    -3
    -4     0     9   -15
```

We will illustrate the operation R1 by adding 4 times the first row to the third row. The notation for the third row is M(3,:), so we will replace this by M(3,:) + 4*M(1,:). We will let MATLAB do all of the tedious arithmetic.

```
>> M(3,:) = M(3,:)+4*M(1,:)
M =

     1     3     5     6
     3     5     0    -3
     0    12    29     9
```

To illustrate operation R2 we will exchange rows 2 and 3. In this case we want M([2,3],:) to be equal to what M([3,2],:) is currently, so we execute the following command.

```
>> M([2,3],:)=M([3,2],:)
M =

     1     3     5     6
     0    12    29     9
     3     5     0    -3
```

Finally, we illustrate R3 by multiplying row 2 by -5.

```
>> M(2,:) = -5*M(2,:)
M =

     1     3     5     6
     0   -60  -145   -45
     3     5     0    -3
```

We can divide a row by a number just as easily

```
>> M(3,:) = M(3,:)/3
M =

    1.0000    3.0000    5.0000    6.0000
         0  -60.0000 -145.0000  -45.0000
    1.0000    1.6667         0   -1.0000
```

146

The Rational Format

Frequently problems arise in which the numbers we are using are all rational numbers. The floating point arithmetic that MATLAB uses does not lend itself to maintaining the rational form of numbers. We can sometimes recognize a rational number from the decimal expansions MATLAB gives us, but frequently, we can't. For example, in M as it appears above, we know that M(3,2)=5/3, but we will not always be so fortunate. MATLAB provides a solution with its rational format. If we enter `format rat` at the MATLAB prompt, from that point on MATLAB will display numbers as the ratios of integers. For example,

```
>> format rat
>> pi
ans =
   355/113
```

We know, and MATLAB knows, that $\pi \neq 355/113$. MATLAB is only displaying an approximation to the answer. But this is no different than what MATLAB always does. The only difference is that now MATLAB is using a rational number to approximate π instead of a decimal expansion. It computes and displays the rational number that is closest to the given number within a prescribed tolerance. The advantage is that if we know the answer is a rational number, then we know that the rational representation shown us by MATLAB is probably 100% correct.

If we display our current matrix M, now we see the following:

```
>> M
M =
    1             3             5             6
    0           -60          -145           -45
    1           5/3            0            -1
```

Thus we get the precise, rational version of M.

There are some things about the rational format that can be confusing. We will inevitably run into them as we proceed. For one thing, asterisks (∗) sometimes appear in the output. These appear when MATLAB is trying to give a rational approximation to a very small number. This will require a very large denominator, and the resulting expression will not fit within the rational format. When we are dealing with what we know to be rational numbers with small denominators, these large denominators that appear are due to round off error. They usually represent a number which actually should be 0, and they can usually be replaced by 0, but caution is called for in doing so.

Solving Linear Equations

In this section we set ourselves the task of solving systems of linear equations. First, let's pause to define a matrix form called *reduced row echelon form*. In general a matrix is said to be in reduced row echelon form if it has the form

$$\begin{bmatrix} 1 & 0 & \# & \# & 0 & 0 & \# & \# & 0 & \# \\ 0 & 1 & \# & \# & 0 & 0 & \# & \# & 0 & \# \\ 0 & 0 & 0 & 0 & 1 & 0 & \# & \# & 0 & \# \\ 0 & 0 & 0 & 0 & 0 & 1 & \# & \# & 0 & \# \\ 0 & 0 & 0 & 0 & 0 & 0 & 0 & 0 & 1 & \# \\ 0 & 0 & 0 & 0 & 0 & 0 & 0 & 0 & 0 & 0 \\ 0 & 0 & 0 & 0 & 0 & 0 & 0 & 0 & 0 & 0 \end{bmatrix},$$

where the #'s stand for arbitrary numbers. A more formal definition follows.

First we define the *pivot* of a row to be the first nonzero entry in that row.

Reduced Row Echelon Form. *A matrix is in reduced row echelon form if it has the following properties:*

- *There may be rows consisting entirely of zeros (i.e., rows without pivots), but these rows are collected at the bottom of the matrix.*
- *Each pivot is a 1.*
- *Each pivot is in a column strictly to the right of the pivot in the rows above.*

Our entire strategy for solving systems of linear equations amounts to taking the augmented matrix of the system and transforming it to reduced row echelon form using elementary row operations. This method is called *Gaussian elimination*.

We will call a column which contains a pivot a *pivot column*. A column which does not contain a pivot is called a *free column*. As we will see, the free columns correspond to *free variables*, i.e., variables to which we are free to assign arbitrary values in the solution of our equations.

Example 1. *Solve the system*

$$\begin{aligned} 3x_1 - 4x_2 + 5x_3 - x_4 &= 5, \\ x_2 - x_3 + 6x_4 &= 0, \\ 5x_1 \quad\quad - x_3 + 4x_4 &= 4, \end{aligned} \tag{9.4}$$

for x_1, x_2, x_3, and x_4.

Write system (9.4) as a matrix equation.

$$\begin{bmatrix} 3 & -4 & 5 & -1 \\ 0 & 1 & -1 & 6 \\ 5 & 0 & -1 & 4 \end{bmatrix} \begin{bmatrix} x_1 \\ x_2 \\ x_3 \\ x_4 \end{bmatrix} = \begin{bmatrix} 5 \\ 0 \\ 4 \end{bmatrix} \tag{9.5}$$

The matrix equation (9.5) has the form $A\mathbf{x} = \mathbf{b}$, where

$$A = \begin{bmatrix} 3 & -4 & 5 & -1 \\ 0 & 1 & -1 & 6 \\ 5 & 0 & -1 & 4 \end{bmatrix}, \quad \mathbf{x} = \begin{bmatrix} x_1 \\ x_2 \\ x_3 \\ x_4 \end{bmatrix}, \quad \text{and} \quad \mathbf{b} = \begin{bmatrix} 5 \\ 0 \\ 4 \end{bmatrix}.$$

In MATLAB, we proceed as follows:

```
>> A=[3 -4 5 -1;0 1 -1 6;5 0 -1 4]
A =
     3          -4          5          -1
     0           1         -1           6
     5           0         -1           4

>> b=[5 0 4]'
b =
     5
     0
     4
```

To analyze the system using row operations, we augment the matrix A by adding the vector b as a fifth column. We will denote the augmented matrix by M.

```
>> M=[A,b]
M =
     3          -4          5          -1           5
     0           1         -1           6           0
     5           0         -1           4           4
```

Now we transform M by row operations to get an equivalent system which is easier to solve. We start by making $M(1,1)=1$. To do this we divide row 1 by 3:

```
>> M(1,:)=M(1,:)/3
M =
     1        -4/3         5/3        -1/3         5/3
     0           1          -1           6           0
     5           0          -1           4           4
```

Notice that we are using the rational format. Next we eliminate the non-zero entries in the first column by adding or subtracting appropriate multiples of the first row.

```
>> M(3,:)=M(3,:)-5*M(1,:)
M =
     1        -4/3         5/3        -1/3         5/3
     0           1          -1           6           0
     0        20/3       -28/3        17/3       -13/3
```

Note that the pivot in $M(2,2)$ is already a 1. We use this pivot to eliminate non-zero entries in the second column by adding multiples of the second row to the first and the third rows.

149

```
>> M(1,:)=M(1,:)+(4/3)*M(2,:)
M =
        1          0          1/3        23/3        5/3
        0          1          -1         6           0
        0          20/3       -28/3      17/3        -13/3

>> M(3,:)=M(3,:)-(20/3)*M(2,:)
M =
        1          0          1/3        23/3        5/3
        0          1          -1         6           0
        0          *          -8/3       -103/3      -13/3
```

That asterisk (*) in the bottom row should be a zero. To find its value, enter

```
>> M(3,2)
ans =
    -1/1125899906842624
```

The asterisk indicates that the rational approximation of the number is too large to fit into the format of the matrix. But M(3,2) should be precisely zero, not almost zero as it is here. We have been a little careless. When we do row operations, we should refer to the matrix elements by name instead of trying to guess the number that MATLAB is using for them. Remember, MATLAB is dealing with approximations all the time. It is constantly making very small mistakes due to round off error. To fix this we proceed:

```
>> M(3,:)=M(3,:)-M(3,2)*M(2,:)
M =
        1          0          1/3        23/3        5/3
        0          1          -1         6           0
        0          0          -8/3       -103/3      -13/3
```

Notice that we referred directly to M(3,2) in this command. This is the preferred method for doing row operations. We will use this procedure in the future, and we recommend that you do too.

Next we divide the third row by M(3,3) to make that element 1.

```
>> M(3,:)=M(3,:)/M(3,3)
M =
        1          0          1/3        23/3        5/3
        0          1          -1         6           0
        0          0          1          103/8       13/8
```

Use the pivot established in M(3,3) to eliminate remaining non-zero entries in the third column by adding appropriate multiples of the third row to the first and second rows.

```
>> M(1,:)=M(1,:)-M(1,3)*M(3,:);
>> M(2,:)=M(2,:)-M(2,3)*M(3,:)
M =
        1          0          0          27/8        9/8
        0          1          0          151/8       13/8
        0          0          1          103/8       13/8
```

150

This last matrix M is in reduced row echelon form. Note the pivots in columns 1, 2, and 3. These columns are the pivot columns. Column 4 is a free column. Consequently, the variables x_1, x_2, and x_3 are pivot variables, while the variable x_4 is a free variable. The matrix M is equivalent to the system

$$x_1 + (27/8)x_4 = 9/8,$$
$$x_2 + (151/8)x_4 = 13/8,$$
$$x_3 + (103/8)x_4 = 13/8.$$

Solve each equation of this system for pivot variables x_1, x_2, and x_3 in terms of the free variable x_4. If we let $x_4 = t$, where t is an arbitrary number, then the solution is

$$x_1 = 9/8 - (27/8)t,$$
$$x_2 = 13/8 - (151/8)t,$$
$$x_3 = 13/8 - (103/8)t,$$
$$x_4 = t,$$

where t is any arbitrary number. Of course, the elementary row operations yield an equivalent system, so this solution is also the solution of system (9.4).

This solution is sometimes called a *general* solution, as all possible solutions of system (9.4) can be garnered by substituting arbitrary numbers for the parameter t. For example, $t = 0$ yields the solution $x_1 = 9/8$, $x_2 = 13/8$, $x_3 = 13/8$, and $x_4 = 0$; $t = 1$ yields the solution $x_1 = -18/8$, $x_2 = -138/8$, $x_3 = -90/8$, and $x_4 = 1$; etc.

After completing this example, the reader must be wondering why the computer and MATLAB can't do the whole process at once. In fact it can. If we reform the original augmented matrix M,

```
>> M=[A,b]
M =
     3    -4     5    -1     5
     0     1    -1     6     0
     5     0    -1     4     4,
```

then we can proceed directly to the reduced row echelon form by using the MATLAB command `rref`.

```
>> MR=rref(M)
MR =
     1         0         0      27/8       9/8
     0         1         0     151/8      13/8
     0         0         1     103/8      13/8
```

This eliminates all of the intermediate steps, and we can immediately write down the equivalent system of equations as we did earlier.

Example 2. *Solve the system*

$$2x_2 + 2x_3 + 3x_4 = -4,$$
$$-2x_1 + 4x_2 + 2x_3 - x_4 = -6, \qquad (9.6)$$
$$3x_1 - 4x_2 - x_3 + 2x_4 = 8.$$

151

The augmented matrix is

```
>> M = [0 2 2 3 -4; -2 4 2 -1 -6; 3 -4 -1 2 8]
M =
         0         2         2         3        -4
        -2         4         2        -1        -6
         3        -4        -1         2         8
```

Place the augmented matrix in reduced row echelon form as follows:

```
>> rref(M)
ans =
         1         0         1         0      16/5
         0         1         1         0      -1/5
         0         0         0         1      -6/5
```

Note that the pivot columns are one, two, and four, making x_1, x_2, and x_4 pivot variables, leaving x_3 as a free variable. The equivalent system is

$$x_1 + x_3 = 16/5,$$
$$x_2 + x_3 = -1/5,$$
$$x_4 = -6/5.$$

Solve each of the above equations for a pivot variable, in terms of the free variable x_3.

$$x_1 = 16/5 - x_3$$
$$x_2 = -1/5 - x_3$$
$$x_4 = -6/5$$

Set $x_3 = t$, where t is an arbitrary number, but this time let's write the solution in vector form.

$$\mathbf{x} = \begin{bmatrix} x_1 \\ x_2 \\ x_3 \\ x_4 \end{bmatrix} = \begin{bmatrix} 16/5 - t \\ -1/5 - t \\ t \\ -6/5 \end{bmatrix} = \begin{bmatrix} 16/5 \\ -1/5 \\ 0 \\ -6/5 \end{bmatrix} + t \begin{bmatrix} -1 \\ -1 \\ 1 \\ 0 \end{bmatrix}.$$

Determined Systems of Equations

A system is *determined* if there are the same number of equations and unknowns. It is *underdetermined* if there are fewer equations than unknowns, and *overdetermined* if there are more equations than unknowns. The systems we have been solving up to now have been underdetermined. In this section we examine determined systems.

There are essentially three things that can happen when solving a system: the system has no solutions, the system has an infinite number of solutions, or the system has a unique solution.

Example 3. *Solve the system*

$$4x_1 - 4x_2 - 8x_3 = 27,$$
$$2x_2 + 2x_3 = -6, \tag{9.7}$$
$$x_1 - 2x_2 - 3x_3 = 10.$$

152

Note that system (9.7) is determined. It has three unknowns (x_1, x_2, and x_3) and three equations. Set up the augmented matrix

```
>> M=[4 -4 -8 27;0 2 2 -6;1 -2 -3 10]
M =
     4        -4        -8        27
     0         2         2        -6
     1        -2        -3        10
```

and place it in reduced row echelon form.

```
>> rref(M)
ans =
     1         0        -1         0
     0         1         1         0
     0         0         0         1
```

Note that the last row of the reduced row echelon form represents the equation $0x_1 + 0x_2 + 0x_3 = 1$, which clearly has no solutions. If there is no solution which satisfies this last equation, then there certainly cannot be a solution which satisfies all three equations represented by the matrix. Consequently, the system (9.7) has no solutions and is called an *inconsistent* system.

Example 4. *Solve the system*

$$4x_1 - 4x_2 - 8x_3 = 4,$$
$$2x_2 + 2x_3 = 2, \tag{9.8}$$
$$x_1 - 2x_2 - 3x_3 = 0.$$

Note that system (9.8) is determined. It has three unknowns (x_1, x_2, and x_3) and three equations. Set up the augmented matrix

```
>> M=[4 -4 -8 4;0 2 2 2;1 -2 -3 0]
M =
     4    -4    -8     4
     0     2     2     2
     1    -2    -3     0
```

and place it in reduced row echelon form.

```
>> rref(M)
ans =
     1     0    -1     2
     0     1     1     1
     0     0     0     0
```

Note that the pivot columns are one and two, leaving x_3 as a free variable. The equivalent system is

$$x_1 - x_3 = 2,$$
$$x_2 + x_3 = 1. \tag{9.9}$$

153

Note that the last row of the matrix, $[0 \ 0 \ 0 \ 0]$, represents the equation $0x_1 + 0x_2 + 0x_3 = 0$. Since any combination of x_1, x_2, and x_3 is a solution of this equation, we need only turn our attention to finding solutions of system (9.9). Solve each equation of system (9.9) for a pivot variable in terms of the free variable x_3.

$$x_1 = 2 + x_3$$
$$x_2 = 1 - x_3$$

If we set $x_3 = t$, where t is an arbitrary number, and place the solution in in vector form, then

$$\mathbf{x} = \begin{bmatrix} x_1 \\ x_2 \\ x_3 \end{bmatrix}, = \begin{bmatrix} 2+t \\ 1-t \\ t \end{bmatrix}, = \begin{bmatrix} 2 \\ 1 \\ 0 \end{bmatrix} + t \begin{bmatrix} 1 \\ -1 \\ 1 \end{bmatrix},$$

where t is any arbitrary number. System (9.8) has an infinite number of solutions.

Example 5. *Solve the system*

$$3x_1 - 4x_2 - 8x_3 = 27,$$
$$2x_2 + 2x_3 = -6, \qquad\qquad (9.10)$$
$$x_1 - 2x_2 - 3x_3 = 10.$$

Enter the augmented matrix

```
>> M=[3 -4 -8 27;0 2 2 -6;1 -2 -3 10]
M =
     3    -4    -8    27
     0     2     2    -6
     1    -2    -3    10
```

and place it in reduced row echelon form.

```
>> rref(M)
ans =
     1     0     0     1
     0     1     0     0
     0     0     1    -3
```

Hence, the solution of system (9.10) is unique.

$$x_1 = 1$$
$$x_2 = 0$$
$$x_3 = -3$$

These examples show that the situation with determined systems is not so simple. Sometimes there are solutions, sometimes not. Sometimes solutions are unique, sometimes not. We need a way to tell which phenomena occur in specific cases. The answers are provided by the *determinant*.

154

The Determinant and Systems

We will not study the determinant in any detail. Let's just recall some facts. The determinant is defined for any square matrix, i.e. any matrix which has the same number of rows and columns. For two by two and three by three matrices we have the formulas

$$\det \begin{bmatrix} a & b \\ c & d \end{bmatrix} = ad - bc$$

and

$$\det \begin{bmatrix} a & b & c \\ d & e & f \\ g & h & i \end{bmatrix} = aei - afh - bdi + cdh + bfg - ceg.$$

For larger matrices the formula for the determinant gets increasingly lengthy, and decreasingly useful for calculation. Once more MATLAB comes to our rescue, since it has a built-in procedure det for calculating the determinants of matrices of arbitrary size.

Consider the determinant of the coefficient matrix of systems (9.7) and (9.8) of Examples 3 and 4.

```
>> A=[4 -4 -8;0 2 2;1 -2 -3]
A =
     4    -4    -8
     0     2     2
     1    -2    -3
>> det(A)
ans =
     0
```

Note that the determinant is zero. On the contrary, note the determinant of the coefficient matrix for system (9.10) of Example 5.

```
>> A=[3 -4 -8;0 2 2;1 -2 -3]
A =
     3    -4    -8
     0     2     2
     1    -2    -3
>> det(A)
ans =
     2
```

The fact that the determinant of the coefficient matrix of system (9.10) is nonzero and the solution is unique is no coincidence.

Theorem 2.

Let A be an $n \times n$ matrix.

a) If $\det A \neq 0$, then for every $n \times 1$ vector $\mathbf{b}$ there is a unique vector $\mathbf{x}$ such that $A\mathbf{x} = \mathbf{b}$. In particular, the only solution to the homogeneous equation $A\mathbf{x} = \mathbf{0}$ is the zero vector, $\mathbf{x} = \mathbf{0}$.

b) If $\det(A) = 0$, then there are vectors $\mathbf{b}$ for which there are no vectors $\mathbf{x}$, with $A\mathbf{x} = \mathbf{b}$. If there is a solution to the equation $A\mathbf{x} = \mathbf{b}$, then there are infinitely many. In particular, the homogeneous equation $A\mathbf{x} = \mathbf{0}$ has nonzero solutions.

If det $A = 0$, we will say that the matrix A is *singular*. As we saw in Examples 3 and 4, if the coefficient matrix is singular, then two possibilities exist: the system $A\mathbf{x} = \mathbf{b}$ has no solutions, or it has an infinite number of solutions.

On the other hand, if det $A \neq 0$, we will say that A is *nonsingular*. If the matrix A is nonsingular, then the system $A\mathbf{x} = \mathbf{b}$ has a unique solution. MATLAB has a convenient way to do this. It is only necessary to enter x=A\b. Hence, system (9.10) can be solved as follows:

```
>> A=[3 -4 -8;0 2 2;1 -2 -3];b=[27;-6;10];
>> x=A\b
x =
        1
        1/1125899906842624
       -3
```

The very large denominator in the second term indicates that MATLAB is approximating 0. (Look at this answer in `format long`.) Thus, this solution agrees with the one found in Example 5. On the other hand, if we try the same technique on system (9.7) of Example 3

```
>> A=[4 -4 -8;0 2 2;1 -2 -3];b=[27;-6;10];
>> x=A\b

Warning: Matrix is singular to working precision.
x =
       1/0
       1/0
       1/0
```

we see that MATLAB recognizes the fact that A is a singular matrix.

The backslash command A\b is deceptively simple. It looks as if we are dividing the equation $A\mathbf{x} = \mathbf{b}$ on the left by the matrix A to produce the solution x=A\b. In a way this is true, but in order to implement that division, MATLAB has to go through a series of computations involving row operations which is very similar to those we discussed earlier in this chapter.

If the matrix A is nonsingular, then it has an *inverse*. This is another matrix B such that $AB = BA = I$. Here I is the *identity matrix*, which has all ones along the diagonal, and zeros off the diagonal. In general it is time-consuming to calculate the inverse of a matrix, but again MATLAB can do this for us. We enter B=inv(A). For example, if A is the coefficient matrix of system (9.10), then

```
>> A=[3 -4 -8;0 2 2;1 -2 -3]
A =
        3           -4           -8
        0            2            2
        1           -2           -3
>> B=inv(A)
B =
       -1            2            4
        1           -1/2         -3
       -1            1            3
```

You can verify that $BA = AB = I$, with B*A and A*B. Try it! By the way, in MATLAB the command eye(n) is used to designate the $n \times n$ identity matrix. Try eye(3).

The Nullspace of a Matrix

The *nullspace* of a matrix A is the set of all vectors x such that $Ax = 0$. Because $x = 0$ is always a solution of $Ax = 0$, the zero vector is always an element of the nullspace of matrix A. However, if A is nonsingular, then Theorem 2 guarantees that this solution is unique, and the nullspace of A consists of only the zero vector 0.

On the other hand, if matrix A is singular, then the nullspace will contain lots of non-zero vectors.

Example 6. *Find the nullspace of matrix*

$$A = \begin{bmatrix} 12 & -8 & -4 & 30 \\ 3 & -2 & -1 & 6 \\ 18 & -12 & -6 & 42 \\ 6 & -4 & -2 & 15 \end{bmatrix}.$$

The nullspace of A consists of all solutions of $Ax = 0$, or

$$\begin{bmatrix} 12 & -8 & -4 & 30 \\ 3 & -2 & -1 & 6 \\ 18 & -12 & -6 & 42 \\ 6 & -4 & -2 & 15 \end{bmatrix} \begin{bmatrix} x_1 \\ x_2 \\ x_3 \\ x_4 \end{bmatrix} = \begin{bmatrix} 0 \\ 0 \\ 0 \\ 0 \end{bmatrix}. \tag{9.11}$$

Set up the augmented matrix

```
>> A=[12 -8 -4 30;3 -2 -1 6;18 -12 -6 42;6 -4 -2 15];
>> b=zeros(4,1);
>> M=[A,b];
```

and place it in reduced row echelon form.

```
>> rref(M)
ans =
     1      -2/3     -1/3       0        0
     0       0        0         1        0
     0       0        0         0        0
     0       0        0         0        0
```

Note that the pivot columns are one and four, making x_1 and x_4 pivot variables, leaving x_2 and x_3 as free variables. The equivalent system is

$$x_1 - (2/3)x_2 - (1/3)x_3 = 0,$$
$$x_4 = 0. \tag{9.12}$$

Solve each equation of system (9.12) for a pivot variable, in terms of the free variables x_2 and x_3.

$$x_1 = (2/3)x_2 + (1/3)x_3$$
$$x_4 = 0$$

157

Set $x_2 = s$ and $x_3 = t$, where s and t are arbitrary numbers, then write the solution in vector form.

$$\mathbf{x} = \begin{bmatrix} x_1 \\ x_2 \\ x_3 \\ x_4 \end{bmatrix} = \begin{bmatrix} (2/3)s + (1/3)t \\ s \\ t \\ 0 \end{bmatrix} = \begin{bmatrix} (2/3)s \\ s \\ 0 \\ 0 \end{bmatrix} + \begin{bmatrix} (1/3)t \\ 0 \\ t \\ 0 \end{bmatrix} = s \begin{bmatrix} 2/3 \\ 1 \\ 0 \\ 0 \end{bmatrix} + t \begin{bmatrix} 1/3 \\ 0 \\ 1 \\ 0 \end{bmatrix}$$

Thus, every element in the nullspace of matrix A can be written as a linear combination[1] of the vectors $\mathbf{v}_1 = [2/3, 1, 0, 0]^T$ and $\mathbf{v}_2 = [1/3, 0, 1, 0]^T$. The nullspace contains infinitely many vectors, but if we know $\mathbf{v}_1$ and $\mathbf{v}_2$ we know them all.

The results in this example exemplify what happens in general. The nullspace of a matrix will consist of all linear combinations of a small number of explicit vectors. If these vectors are linearly independent (see the next section), they are called a *basis* of the nullspace, and the nullspace is said to be *spanned* by the basis vectors. If we know a basis for a nullspace, then we know all of the vectors in the nullspace. For this reason we will almost always describe a nullspace by giving a basis.

Linear Dependence and Independence

A finite set of vectors is said to be *linearly independent* if the only linear combination which is equal to the zero vector is the one where all of the coefficients are equal to zero; i.e., the vectors $\mathbf{v}_1, \mathbf{v}_2, \cdots, \mathbf{v}_p$ are linearly independent if $c_1\mathbf{v}_1 + c_2\mathbf{v}_2 + \cdots + c_p\mathbf{v}_p = \mathbf{0}$ implies that $c_1 = c_2 = \cdots = c_p = 0$. A set of vectors which is not linearly independent is said to be *linearly dependent*.

Example 7. *Show that the vectors*

$$\mathbf{v}_1 = \begin{bmatrix} 1 \\ 1 \\ 1 \end{bmatrix}, \quad \mathbf{v}_2 = \begin{bmatrix} 0 \\ -1 \\ 2 \end{bmatrix}, \quad and \quad \mathbf{v}_1 = \begin{bmatrix} 2 \\ -1 \\ 8 \end{bmatrix}$$

are linearly dependent.

We need to find a non-trivial solution of

$$c_1 \begin{bmatrix} 1 \\ 1 \\ 1 \end{bmatrix} + c_2 \begin{bmatrix} 0 \\ -1 \\ 2 \end{bmatrix} + c_3 \begin{bmatrix} 2 \\ -1 \\ 8 \end{bmatrix} = \begin{bmatrix} 0 \\ 0 \\ 0 \end{bmatrix}. \tag{9.13}$$

The vector equation (9.13) is equivalent to the matrix equation

$$\begin{bmatrix} 1 & 0 & 2 \\ 1 & -1 & -1 \\ 1 & 2 & 8 \end{bmatrix} \begin{bmatrix} c_1 \\ c_2 \\ c_3 \end{bmatrix} = \begin{bmatrix} 0 \\ 0 \\ 0 \end{bmatrix}. \tag{9.14}$$

Set up the augmented matrix and reduce.

---

[1] If $c_1, c_2, \ldots, c_p$ are numbers and $\mathbf{v}_1, \mathbf{v}_2, \ldots, \mathbf{v}_p$ are vectors, then $c_1\mathbf{v}_1 + c_2\mathbf{v}_2 + \cdots + c_p\mathbf{v}_p$ is called a linear combination of the vectors $\mathbf{v}_1, \mathbf{v}_2, \ldots, \mathbf{v}_p$.

```
>> A=[1 0 2;1 -1 -1;1 2 8];b=zeros(3,1);
>> M=[A,b];
>> rref(M)
ans =
     1        0        2        0
     0        1        3        0
     0        0        0        0
```

The solution is

$$c_1 = -2c_3,$$
$$c_2 = -3c_3,$$

where we are allowed to choose any number we wish for c_3. For example, if we choose $c_3 = 1$, then $c_1 = -2$ and $c_2 = -3$. Readers should check that these numbers satisfy equation (9.13). Consequently, $-2\mathbf{v}_1 - 3\mathbf{v}_2 + 1\mathbf{v}_3 = \mathbf{0}$ and the vectors are dependent.

The Nullspace and Dependence

A quick glance at equation (9.14) reveals that the dependence of the vectors

$$\mathbf{v}_1 = \begin{bmatrix} 1 \\ 1 \\ 1 \end{bmatrix}, \quad \mathbf{v}_2 = \begin{bmatrix} 0 \\ -1 \\ 2 \end{bmatrix}, \quad \text{and} \quad \mathbf{v}_1 = \begin{bmatrix} 2 \\ -1 \\ 8 \end{bmatrix}$$

is related to the nullspace of the matrix

$$V = \begin{bmatrix} 1 & 0 & 2 \\ 1 & -1 & -1 \\ 1 & 2 & 8 \end{bmatrix}.$$

where the columns of matrix V are the vectors $\mathbf{v}_1$, $\mathbf{v}_2$, and $\mathbf{v}_3$. If the nullspace of V contains a nonzero vector (as in Example 7), then the columns of V are linearly dependent, and the elements of the nonzero vector form the coefficients of a non-trivial linear combination of the columns of V which is equal to the zero vector. On the other hand, if the nullspace of V contains only the zero vector, then the columns of V are linearly independent.

What is needed is an easier method for computing a basis for the nullspace of a matrix. MATLAB's null command provides the answer.

```
>> v1=[1;1;1];v2=[0;-1;2];v3=[2;-1;8];
>> V=[v1,v2,v3]
V =
     1        0        2
     1       -1       -1
     1        2        8
>> null(V,'r')
ans =
    -2
    -3
     1
```

MATLAB's `null` command, when used with the `'r'` switch, computes a basis for the nullspace in a manner similar to the technique used in Example 6. You can then find elements of the nullspace by taking all possible linear combinations of the basis vectors provided by MATLAB's `null` command. In this case, elements of the nullspace of V are given by taking multiples of the single basis vector $[-2, -3, 1]^T$. For example, $[-4, -6, 2]$, $[-10, -15, 5]$, and $[1, 3/2, -1/2]$ are also in the nullspace[2] of V.

Let's revisit the matrix

$$A = \begin{bmatrix} 12 & -8 & -4 & 30 \\ 3 & -2 & -1 & 6 \\ 18 & -12 & -6 & 42 \\ 6 & -4 & -2 & 15 \end{bmatrix}$$

of Example 6.

```
>> A=[12 -8 -4 30;3 -2 -1 6;18 -12 -6 42;6 -4 -2 15];
>> null(A,'r')
ans =
       2/3            1/3
        1              0
        0              1
        0              0
```

Look familiar? The vectors $v_1 = [2/3, 1, 0, 0]$ and $v_2 = [1/3, 0, 1, 0]$ form a basis for the nullspace of matrix A. Consequently, elements of the nullspace are formed by taking linear combinations of these vectors, much as we saw in Example 6 when we wrote

$$\mathbf{x} = s \begin{bmatrix} 2/3 \\ 1 \\ 0 \\ 0 \end{bmatrix} + t \begin{bmatrix} 1/3 \\ 0 \\ 1 \\ 0 \end{bmatrix}.$$

Bases are not unique. As long as you have two independent vectors that span the nullspace of matrix A, you've got a basis. For example, it is not difficult to show that all elements of the nullspace of matrix A can also be written as linear combinations of the independent vectors $\mathbf{w}_1 = [2, 3, 0, 0]^T$ and $\mathbf{w}_2 = [1, 0, 3, 0]^T$. Also, consider the output from MATLAB's `null` command, this time without the `'r'` switch.

```
>> null(A)
ans =
    0.5973    -0.0194
    0.7023    -0.4702
    0.3873     0.8823
    0.0000     0.0000
```

The output of this `null` command is normalized so that the sum of the squares of the entries of each column vector equals one, and the dot product of different column vectors equals zero. This basis is called an *orthonormal* basis and is very useful in certain situations.

---

[2] And $-4v_1 - 6v_2 + 2v_2 = 0$, etc.

Although bases are not unique, it is true that every basis for the nullspace must contain the same number of vectors. This number is called the *dimension* of the nullspace. Thus, the nullspace of matrix A has dimension equal to 2; and, should you find another basis for the nullspace of the matrix A, it will also contain two vectors.

Example 8. *Discuss the dependence of the vectors*

$$\mathbf{v}_1 = \begin{bmatrix} 1 \\ -1 \\ 4 \end{bmatrix}, \quad \mathbf{v}_2 = \begin{bmatrix} -2 \\ 4 \\ 5 \end{bmatrix}, \quad and \quad \mathbf{v}_1 = \begin{bmatrix} -2 \\ 10 \\ 43 \end{bmatrix}.$$

```
>> v1=[1;-1;4];v2=[-2;4;5];v3=[-2;10;43];
>> V=[v1,v2,v3]
V =
     1        -2        -2
    -1         4        10
     4         5        43
>> null(V,'r')
ans =
   Empty matrix: 3-by-0
```

This response means that there are no nonzero vectors in the nullspace, so we conclude that the vectors are linearly independent.

Exercises

1. Find the general solution to each of the following systems of linear equations using the method of row operations. You may use MATLAB to perform the operations, but in your submission show all of the operations that you perform. In other words, do not use `rref`. Use the `diary` command (as explained in Chapter 1) to record your work.

   a) $-5x + 14y = -47$
      $-7x + 16y = -55$

   b) $2x_1 - 5x_2 + 3x_3 = 8$
      $4x_1 + 3x_2 - 7x_3 = -3$

   c) $-6x - 8y + 8z = -30$
      $9x + 11y - 8z = 33$
      $9x + 9y - 6z = 27$

   d) $-19x_1 - 128x_2 + 81x_3 + 38x_4 = 0$
      $8x_1 + 61x_2 - 27x_3 - 16x_4 = 0$
      $2x_1 + 4x_2 + 12x_3 - 4x_4 = 0$
      $-8x_1 - 16x_2 + 12x_3 + 16x_4 = 0$

2. Find the general solution to each of the following systems of linear equations. In this problem you may use `rref`. You may also use the backslash operation (A\b) if it applies.

   a) $-x - 9y + 6z = 15$
      $2x + 9y + z = -16$

   b) $8x - 10y - 228z = -112$
      $2x - y - 4z = -16$
      $4x - 5y - 14z = -56$

   c) $-19x_1 - 128x_2 + 81x_3 + 38x_4 = 3$
      $8x_1 + 61x_2 - 27x_3 - 16x_4 = 5$
      $2x_1 + 4x_2 + 12x_3 - 4x_4 = -21$
      $-8x_1 - 16x_2 + 12x_3 + 16x_4 = -4$

   d) $2x + 2z = 6$
      $x + z = 3$
      $-7x + 12y + 5z = 27$

161

e) $\quad \frac{2}{3}x_1 \quad - \frac{2}{3}x_3 + \frac{1}{3}x_4 = \frac{1}{3}$

$\quad \frac{11}{2}x_1 - 3x_2 - \frac{20}{3}x_3 + \frac{1}{3}x_4 = -\frac{80}{3}$

$\quad -3x_1 + 3x_2 + 6x_3 \quad\quad = 27$

$\quad -\frac{16}{3}x_1 + 6x_2 + \frac{34}{3}x_3 + \frac{1}{3}x_4 = \frac{163}{3}$

f) $\quad -23x_1 + 26x_2 - 42x_3 - 32x_4 - 90x_5 = -6$

$\quad -2x_1 + \quad x_2 \quad\quad - 3x_4 - 4x_5 = -2$

$\quad -17x_1 + 19x_2 - 28x_3 - 22x_4 - 63x_5 = -3$

$\quad -14x_1 + 14x_2 - 24x_3 - 16x_4 - 52x_5 = -2$

$\quad 18x_1 - 20x_2 + 32x_3 + 23x_4 + 69x_5 = 3$

3. For each of the following matrices find a basis for the nullspace. What is the dimension in each case?

a) $\begin{bmatrix} 2 & 2 \\ -1 & -1 \end{bmatrix}$

b) $\begin{bmatrix} 0 & 4 & -2 \\ 1 & -2 & 1 \\ 3 & -10 & 5 \end{bmatrix}$

c) $\begin{bmatrix} 2 & -4 & -9 \\ 0 & 0 & -2 \\ 0 & 0 & 1 \end{bmatrix}$

d) $\begin{bmatrix} 12 & -5 & -14 & -9 \\ 18 & -7 & -22 & -12 \\ 12 & -6 & -12 & -9 \\ -16 & 8 & 16 & 11 \end{bmatrix}$

e) $\begin{bmatrix} -3 & 2 & 5 & 2 \\ 6 & -2 & -8 & -2 \\ -4 & 2 & 6 & 2 \\ -4 & 2 & 6 & 2 \end{bmatrix}$

f) $\begin{bmatrix} 6 & -4 & 2 & -12 & -8 \\ 6 & -4 & 2 & -12 & -8 \\ 29 & -14 & 11 & -54 & -36 \\ 13 & -6 & 5 & -24 & -16 \\ -10 & 4 & -4 & 18 & 12 \end{bmatrix}$

4. For each of the following set of vectors, determine if they are linearly dependent or independent. If they are dependent, find a nonzero linear combination which is equal to the zero vector.

a) $\begin{bmatrix} 1 \\ 2 \end{bmatrix}$ and $\begin{bmatrix} -3 \\ -6 \end{bmatrix}$

b) $\begin{bmatrix} 1 \\ 2 \end{bmatrix}$ and $\begin{bmatrix} -3 \\ -5 \end{bmatrix}$

c) $\begin{bmatrix} 1 \\ 2 \\ 0 \end{bmatrix}$ and $\begin{bmatrix} -3 \\ -6 \\ 1 \end{bmatrix}$

d) $\begin{bmatrix} 1 \\ 1 \\ 1 \end{bmatrix}$, $\begin{bmatrix} 1 \\ -1 \\ 1 \end{bmatrix}$, and $\begin{bmatrix} 5 \\ 0 \\ 5 \end{bmatrix}$

e) $\begin{bmatrix} 1 \\ 1 \\ 1 \end{bmatrix}$, $\begin{bmatrix} 1 \\ -1 \\ 1 \end{bmatrix}$, and $\begin{bmatrix} 5 \\ 1 \\ 5 \end{bmatrix}$

f) $\begin{bmatrix} 1 \\ 0 \\ 1 \\ 0 \end{bmatrix}$, $\begin{bmatrix} 0 \\ 1 \\ 1 \\ 1 \end{bmatrix}$, and $\begin{bmatrix} 5 \\ -6 \\ -1 \\ -6 \end{bmatrix}$

g) $\begin{bmatrix} 1 \\ 0 \\ 1 \\ 0 \end{bmatrix}$, $\begin{bmatrix} 0 \\ 1 \\ 1 \\ 1 \end{bmatrix}$, and $\begin{bmatrix} 5 \\ -6 \\ 0 \\ -6 \end{bmatrix}$

10.   Homogeneous Linear Systems of ODEs

A system of first order differential equations is *linear* and *homogeneous*, with *constant* coefficients, if it has the specific form

$$
\begin{aligned}
x_1' &= a_{11}x_1 + a_{12}x_2 + \cdots + a_{1n}x_n, \\
x_2' &= a_{21}x_1 + a_{22}x_2 + \cdots + a_{2n}x_n, \\
&\ \ \vdots \\
x_n' &= a_{n1}x_1 + a_{n2}x_2 + \cdots + a_{nn}x_n,
\end{aligned}
\tag{10.1}
$$

where each a_{ij} is a constant number. System (10.1) can be written in matrix form

$$
\begin{bmatrix} x_1 \\ x_2 \\ \vdots \\ x_n \end{bmatrix}' =
\begin{bmatrix} a_{11} & a_{12} & \cdots & a_{1n} \\ a_{21} & a_{22} & \cdots & a_{2n} \\ \vdots & \vdots & \ddots & \vdots \\ a_{n1} & a_{n2} & \cdots & a_{nn} \end{bmatrix}
\begin{bmatrix} x_1 \\ x_2 \\ \vdots \\ x_n \end{bmatrix},
\tag{10.2}
$$

and more compactly, as

$$
\mathbf{x}' = A\mathbf{x}.
\tag{10.3}
$$

Note that equation (10.3) is strikingly similar to the simple first order differential equation

$$
x' = \lambda x.
\tag{10.4}
$$

It is easy to check that

$$
x = Ce^{\lambda t}
\tag{10.5}
$$

is a solution to equation (10.4), where C is an arbitrary constant. Consequently, it seems natural that we should first try to find a solution of equation (10.3) having the form

$$
\mathbf{x} = e^{\lambda t}\mathbf{v},
\tag{10.6}
$$

where $\mathbf{v}$ is a $n \times 1$ vector of constants.

Note that $\mathbf{x} = \mathbf{0}$ is a solution of equation (10.3), but not a very interesting one. Let's agree that we want to find non-zero solutions of equation (10.3). That is, we wish to find a solution of the form $\mathbf{x} = e^{\lambda t}\mathbf{v}$, where $\mathbf{v}$ is a non-zero vector of constants (some of $\mathbf{v}$'s entries may be zero, but not all). If we substitute equation (10.6) into the equation (10.3), then

$$
\begin{aligned}
\mathbf{x}' &= A\mathbf{x}, \\
\left(e^{\lambda t}\mathbf{v}\right)' &= A\left(e^{\lambda t}\mathbf{v}\right), \\
\lambda e^{\lambda t}\mathbf{v} &= e^{\lambda t}A\mathbf{v}.
\end{aligned}
$$

If we divide both sides of this last equation by $e^{\lambda t}$, which is non-zero, we arrive at

$$
A\mathbf{v} = \lambda\mathbf{v}.
\tag{10.7}
$$

The problem of finding a solution of equation (10.3) is now reduced to finding a number λ and a non-zero vector of constants $\mathbf{v}$ that satisfy the equation $A\mathbf{v} = \lambda\mathbf{v}$.

Eigenvalues Using MATLAB

Solutions of the equation $A\mathbf{v} = \lambda\mathbf{v}$, though important for the solution of equation (10.3), have far reaching application in many areas of mathematics and science. Consequently, we pause to make a definition.

Definition. *Let A be an $n \times n$ matrix. A number λ is said to be an eigenvalue for A if there is a non-zero vector $\mathbf{v}$ such that $A\mathbf{v} = \lambda\mathbf{v}$. If λ is an eigenvalue for A, then any vector $\mathbf{v}$ which satisfies $A\mathbf{v} = \lambda\mathbf{v}$ is called an eigenvector for A associated to the eigenvalue λ.*

Let I be an $n \times n$ identity matrix with ones along the main diagonal and zeros elsewhere. If $\mathbf{v}$ is an $n \times 1$ vector, then $I\mathbf{v} = \mathbf{v}$. We can now manipulate the equation $A\mathbf{v} = \lambda\mathbf{v}$ as follows:

$$0 = A\mathbf{v} - \lambda\mathbf{v},$$
$$0 = A\mathbf{v} - \lambda I\mathbf{v},$$
$$0 = (A - \lambda I)\mathbf{v}. \tag{10.8}$$

We now apply Theorem 2 of Chapter 9. If $\det(A - \lambda I) \neq 0$, then equation (10.8) has the *unique* solution $\mathbf{v} = \mathbf{0}$. However, we are looking for *non-zero* solutions of equation (10.8), so we conclude that $\det(A - \lambda I)$ must equal zero.

Definition. *Let A be an $n \times n$ matrix. The function*

$$p(\lambda) = \det(A - \lambda I)$$

is called the **characteristic polynomial** *of the matrix A. The zeros of the characteristic polynomial are the eigenvalues of the matrix A.*

Example 1. *Find the eigenvalues of the matrix*

$$A = \begin{bmatrix} -3 & 1 & -3 \\ -8 & 3 & -6 \\ 2 & -1 & 2 \end{bmatrix}.$$

Calculate the characteristic polynomial as follows:

$$p(\lambda) = \det(A - \lambda I),$$
$$= \det\left(\begin{bmatrix} -3 & 1 & -3 \\ -8 & 3 & -6 \\ 2 & -1 & 2 \end{bmatrix} - \lambda \begin{bmatrix} 1 & 0 & 0 \\ 0 & 1 & 0 \\ 0 & 0 & 1 \end{bmatrix}\right),$$
$$= \det\begin{bmatrix} -3-\lambda & 1 & -3 \\ -8 & 3-\lambda & -6 \\ 2 & -1 & 2-\lambda \end{bmatrix},$$
$$= \lambda^3 - 2\lambda^2 - \lambda + 2.$$

Of course, this last step requires that you are pretty good at calculating the determinant of a rather complicated matrix. If not, let MATLAB assist you with the computation.

164

```
>> A=[-3 1 -3;-8 3 -6;2 -1 2];
>> p = poly(A)
p =
        1            -2           -1            2
```

Note that the vector p=[1 -2 -1 2] contains the coefficients of the characteristic polynomial $p(\lambda) = \lambda^3 - 2\lambda^2 - \lambda + 2$.

To find the eigenvalues we must find the zeros of the characteristic polynomial.

$$p(\lambda) = \lambda^3 - 2\lambda^2 - \lambda + 2 = (\lambda - 1)(\lambda + 1)(\lambda - 2)$$

Thus, the eigenvalues of A are $\lambda = 1, -1$, and 2. Of course, you must be pretty good at factoring to complete the computation using this approach. If not, MATLAB can assist you and even save you some time. If the vector p contains the coefficients of the characteristic polynomial, then we can use roots(p) to compute the zeros of p.

```
>> roots(p)
ans =
        2
        1
       -1
```

Symbolic Toolbox Users. If you change your matrix into a symbolic object, then Maple will be called upon to perform the evaluations. For example,

```
>> ps=poly(sym(A))
ps =
x^3-2*x^2-x+2
```

computes the characteristic polynomial and

```
>> solve(ps)
ans =
[ 1]
[ 2]
[-1]
```

will find the zeros of the characteristic polynomial. The Symbolic Toolbox has another command, factor, which is also of interest here. The command

```
>> factor(ps)
ans =
(x-1)*(x-2)*(x+1)
```

factors the characteristic polynomial stored in ps.

Eigenvectors Using MATLAB

Having found the eigenvalues, we must now find the associated eigenvectors. Equation (10.8), repeated here for emphasis,

$$\mathbf{0} = (A - \lambda I)\,\mathbf{v}, \tag{10.8}$$

is used to find the eigenvector associated with the eigenvalue λ. The eigenvector $\mathbf{v}$, being a solution of $\mathbf{0} = (A - \lambda I)\,\mathbf{v}$, is therefore an element of the nullspace of the matrix $A - \lambda I$.

Example 2. *Find the eigenvectors of the matrix*

$$A = \begin{bmatrix} -3 & 1 & -3 \\ -8 & 3 & -6 \\ 2 & -1 & 2 \end{bmatrix}.$$

The eigenvalues of matrix A were determined in Example 1 to be $\lambda = 1, -1$, and 2. Select one of the eigenvalues, say $\lambda_1 = 1$. The eigenvector associated with $\lambda_1 = 1$ is an element of the nullspace of $A - 1I$ and is found with the following command. (Note that the matrix I is a 3×3 identity matrix, generated with the MATLAB command eye(3).)

```
>> v1=null(A-1*eye(3),'r')
v1 =
    -1
    -1
     1
```

Any vector of an eigenspace can serve as an eigenvector, so let's choose $\mathbf{v}_1 = [-1, -1, 1]^T$ to be the eigenvalue associated with $\lambda_1 = 1$. In a similar manner, the MATLAB commands

```
>> v2=null(A-(-1)*eye(3),'r')
v2 =
    0.5000
    1.0000
         0
>> v3=null(A-2*eye(3),'r')
v3 =
    -1
    -2
     1
```

provide eigenvectors $\mathbf{v}_2 = [1/2, 1, 0]^T$ and $\mathbf{v}_3 = [-1, -2, 1]^T$ associated with the eigenvalues $\lambda_2 = -1$ and $\lambda_3 = 2$, respectively.

Of course, any scalar multiple of an eigenvector will also be an eigenvector[1], so we could just as easily pair the eigenvector $\mathbf{v}_2 = [1, 2, 0]^T$ with the eigenvalue $\lambda = -1$.

---

[1] If $\mathbf{v}$ is an eigenvector of matrix A associated with the eigenvalue λ, then $A(c\mathbf{v}) = c(A\mathbf{v}) = c(\lambda\mathbf{v}) = \lambda(c\mathbf{v})$, making $c\mathbf{v}$ an eigenvector for any constant number c.

Symbolic Toolbox Users. The `null` command works equally well with symbolic objects, but the `'r'` switch is not needed.

```
>> v2=null(sym(A)-(-1)*eye(3))
v2 =
[ 1]
[ 2]
[ 0]
```

Finally, you can easily check your solution, symbolically (with the Symbolic Toolbox) or numerically (with MATLAB). For example, use the Symbolic Toolbox to show that $A\mathbf{v}_2 = -1\mathbf{v}_2$.

```
>> A*v2,-1*v2
ans =
[ -1]
[ -2]
[  0]
ans =
[ -1]
[ -2]
[  0]
```

Since $A\mathbf{v}_2 = -1\mathbf{v}_2$, $\mathbf{v}_2$ is an eigenvector with associated eigenvalue $\lambda_2 = -1$.

MATLAB's EIG Command

Finding eigenvalues and eigenvectors, as we have described it, requires three steps. First, find the characteristic polynomial; second, find the eigenvalues (which are the roots of the characteristic polynomial); and third, for each eigenvalue λ, find the associated eigenvectors by finding the nullspace of $A - \lambda I$. The third step can be accomplished either by row operations on the matrix $A - \lambda I$, or by using the command `null` applied to $A - \lambda I$.

This three step process can be replaced with a one step process using MATLAB's `eig` command.

Example 3. *Use MATLAB's* `eig` *command to find the eigenvalues and eigenvectors of*

$$A = \begin{bmatrix} -3 & 1 & -3 \\ -8 & 3 & -6 \\ 2 & -1 & 2 \end{bmatrix}.$$

The command `eig(A)` will display the eigenvalues of A.

```
>> A=[-3 1 -3;-8 3 -6;2 -1 2];
>> eig(A)
ans =
     2
    -1
     1
```

However, the command [V,E] = eig(A) will output two matrices. E will be a diagonal matrix with the eigenvalues along the diagonal, and V will have the associated eigenvectors as its column vectors.

```
>> [V,E]=eig(A)
V =
     0.4082     0.4472    -0.5774
     0.8165     0.8944    -0.5774
    -0.4082     0.0000     0.5774

E =
     2.0000          0          0
          0    -1.0000          0
          0          0     1.0000
```

The eigenvalue 2 is associated with the eigenvector in the first column of V, the eigenvalue -1 is associated with the eigenvector in the second column of V, and the eigenvalue 1 is associated with the eigenvector in the third column of V.

Eigenvectors are basis elements for the nullspace of $A - \lambda I$. As we saw in Chapter 9, bases are not unique. The eigenvectors that are produced by eig are normalized so that the sum of the squares of the elements of each eigenvector is equal to 1 (the vectors have length one and are *unit vectors*).

You might find the following commands illuminating (or even useful).

```
>> V(:,1)/V(3,1)
ans =
    -1.0000
    -2.0000
     1.0000
```

```
>> V(:,2)/V(2,2)
ans =
     0.5000
     1.0000
     0.0000
```

```
>> V(:,3)/V(3,3)
ans =
    -1.0000
    -1.0000
     1.0000
```

Consequently, each column of the matrix V is a scalar multiple of the corresponding eigenvector produced with the null command in Example 2 (see footnote 1).

Symbolic Toolbox Users. MATLAB's `eig` command will access Maple if the matrix A is a symbolic object.

```
>> [V,E]=eig(sym(A))
V =
[  1, -1, -1]
[  2, -2, -1]
[  0,  1,  1]

E =
[ -1,  0,  0]
[  0,  2,  0]
[  0,  0,  1]
```

These eigenvectors are remarkably similar to those found in Example 2.

Tying It Together—Solving Systems

We are now ready to solve homogeneous, linear systems with constant coefficients.

Example 4. *Solve the the initial value problem*

$$
\begin{aligned}
x_1' &= 8x_1 - 5x_2 + 10x_3, \\
x_2' &= 2x_1 + x_2 + 2x_3, \\
x_3' &= -4x_1 + 4x_2 - 6x_3,
\end{aligned}
\tag{10.9}
$$

where $x_1(0) = 2$, $x_2(0) = 2$, and $x_3(0) = -3$.

Set up the system as a matrix equation.

$$
\begin{bmatrix} x_1 \\ x_2 \\ x_3 \end{bmatrix}' = \begin{bmatrix} 8 & -5 & 10 \\ 2 & 1 & 2 \\ -4 & 4 & -6 \end{bmatrix} \begin{bmatrix} x_1 \\ x_2 \\ x_3 \end{bmatrix}
\tag{10.10}
$$

Equation (10.10) is now in the form $\mathbf{x}' = A\mathbf{x}$. The commands

```
>> A=[8,-5,10;2,1,2;-4,4,-6];
>> eig(A)
```

reveal the eigenvalues $\lambda_1 = -2$, $\lambda_2 = 3$, and $\lambda_3 = 2$, while the commands

```
>> v1=null(A-(-2)*eye(3),'r')
>> v2=null(A-(3)*eye(3),'r')
>> v3=null(A-(2)*eye(3),'r')
```

capture the associated eigenvectors

$$
\mathbf{v}_1 = \begin{bmatrix} -1 \\ 0 \\ 1 \end{bmatrix}, \quad \mathbf{v}_2 = \begin{bmatrix} 1 \\ 1 \\ 0 \end{bmatrix}, \quad \text{and} \quad \mathbf{v}_3 = \begin{bmatrix} 0 \\ 2 \\ 1 \end{bmatrix}.
$$

169

The theory developed at the beginning of the chapter (see equations (10.6) and (10.7)) states that each eigenvalue-eigenvector pair produces a solution $\mathbf{x} = e^{\lambda t}\mathbf{v}$ of $\mathbf{x}' = A\mathbf{x}$.

$$\mathbf{x}_1 = e^{-2t}\begin{bmatrix} -1 \\ 0 \\ 1 \end{bmatrix}, \quad \mathbf{x}_2 = e^{3t}\begin{bmatrix} 1 \\ 1 \\ 0 \end{bmatrix}, \quad \mathbf{x}_3 = e^{2t}\begin{bmatrix} 0 \\ 2 \\ 1 \end{bmatrix}$$

Because the system is linear and the solutions $\mathbf{x}_1$, $\mathbf{x}_2$, and $\mathbf{x}_3$ are linearly independent, the general solution of the system is given by the following linear combination of $\mathbf{x}_1$, $\mathbf{x}_2$, and $\mathbf{x}_3$.

$$\mathbf{x}(t) = c_1 e^{-2t}\begin{bmatrix} -1 \\ 0 \\ 1 \end{bmatrix} + c_2 e^{3t}\begin{bmatrix} 1 \\ 1 \\ 0 \end{bmatrix} + c_3 e^{2t}\begin{bmatrix} 0 \\ 2 \\ 1 \end{bmatrix} \tag{10.11}$$

We need to determine the particular solution that satisfies the initial condition

$$\mathbf{x}(0) = \begin{bmatrix} x_1(0) \\ x_2(0) \\ x_3(0) \end{bmatrix} = \begin{bmatrix} 2 \\ 2 \\ -3 \end{bmatrix}.$$

Substituting $t = 0$ in equation (10.11) yields the following equation.

$$\mathbf{x}(0) = c_1 e^{-2(0)}\begin{bmatrix} -1 \\ 0 \\ 1 \end{bmatrix} + c_2 e^{3(0)}\begin{bmatrix} 1 \\ 1 \\ 0 \end{bmatrix} + c_3 e^{2(0)}\begin{bmatrix} 0 \\ 2 \\ 1 \end{bmatrix}$$

$$\begin{bmatrix} 2 \\ 2 \\ -3 \end{bmatrix} = c_1 \begin{bmatrix} -1 \\ 0 \\ 1 \end{bmatrix} + c_2 \begin{bmatrix} 1 \\ 1 \\ 0 \end{bmatrix} + c_3 \begin{bmatrix} 0 \\ 2 \\ 1 \end{bmatrix} \tag{10.12}$$

Equation (10.12) is equivalent to the matrix equation

$$\begin{bmatrix} -1 & 1 & 0 \\ 0 & 1 & 2 \\ 1 & 0 & 1 \end{bmatrix}\begin{bmatrix} c_1 \\ c_2 \\ c_3 \end{bmatrix} = \begin{bmatrix} 2 \\ 2 \\ -3 \end{bmatrix},$$

an equation in the form $V\mathbf{c} = \mathbf{b}$, whose solution is easily captured with the following MATLAB commands.

```
>> V=[-1 1 0;0 1 2;1 0 1],b=[2;2;-3]
V =
        -1           1           0
         0           1           2
         1           0           1
b =
         2
         2
        -3
>> c=V\b
c =
        -6
        -4
         3
```

Thus, $c_1 = -6$, $c_2 = -4$, $c_3 = 3$, and the solution of the initial value problem (10.9) is

$$\mathbf{x}(t) = -6e^{-2t}\begin{bmatrix} -1 \\ 0 \\ 1 \end{bmatrix} - 4e^{3t}\begin{bmatrix} 1 \\ 1 \\ 0 \end{bmatrix} + 3e^{2t}\begin{bmatrix} 0 \\ 2 \\ 1 \end{bmatrix}.$$

Finally, since $\mathbf{x}(t) = [x_1(t), x_2(t), x_3(t)]^T$, we can write

$$\begin{bmatrix} x_1(t) \\ x_2(t) \\ x_3(t) \end{bmatrix} = \begin{bmatrix} 6e^{-2t} - 4e^{3t} \\ -4e^{3t} + 6e^{2t} \\ -6e^{-2t} + 3e^{2t} \end{bmatrix},$$

yielding the solutions of system (10.9).

Example 5. *Solve the initial value problem*

$$\begin{aligned} x_1' &= -9x_1 + 4x_2, \\ x_2' &= -2x_1 + 2x_2, \end{aligned} \tag{10.13}$$

with initial values $x_1(0) = 1$ and $x_2(0) = -1$.

Set up the system as a matrix equation.

$$\begin{bmatrix} x_1 \\ x_2 \end{bmatrix}' = \begin{bmatrix} -9 & 4 \\ -2 & 2 \end{bmatrix}\begin{bmatrix} x_1 \\ x_2 \end{bmatrix} \tag{10.14}$$

Equation (10.14) is now in the form $\mathbf{x}' = A\mathbf{x}$. The characteristic polynomial of matrix A is $p(\lambda) = \lambda^2 + 7\lambda - 10$, whose zeros are $-7/2 \pm (1/2)\sqrt{89}$. We can give exact values of the eigenvalues and eigenvectors if we are willing to use expressions involving square roots. Instead, let's use the decimal approximations given by MATLAB's eig command.

```
>> A=[-9 4;-2 2];
>> [V,E]=eig(A)
V =
    -0.9814    -0.3646
    -0.1921    -0.9312
E =
    -8.2170         0
         0    1.2170
```

Since scalar multiples of eigenvectors remain eigenvectors, we will still have eigenvectors if we replace V with −V. (Note: we could easily have kept the original eigenvectors. There just were too many negative signs for our taste.)

```
>> V=-V
V =
     0.9814     0.3646
     0.1921     0.9312
```

171

These results easily lead to the general solution.

$$\mathbf{x}(t) = c_1 e^{-8.2170t} \begin{bmatrix} 0.9814 \\ 0.1921 \end{bmatrix} + c_2 e^{1.2170t} \begin{bmatrix} 0.3646 \\ 0.9312 \end{bmatrix} \tag{10.15}$$

We need to determine the solution that satisfies the initial condition

$$\mathbf{x}(0) = \begin{bmatrix} x_1(0) \\ x_2(0) \end{bmatrix} = \begin{bmatrix} 1 \\ -1 \end{bmatrix}.$$

Substituting $t = 0$ in equation (10.15) yields the following equation.

$$\begin{bmatrix} 1 \\ -1 \end{bmatrix} = c_1 \begin{bmatrix} 0.9814 \\ 0.1921 \end{bmatrix} + c_2 \begin{bmatrix} 0.3646 \\ 0.9312 \end{bmatrix} \tag{10.16}$$

Equation (10.16) is equivalent to the matrix equation

$$\begin{bmatrix} 0.9814 & 0.3646 \\ 0.1921 & 0.9312 \end{bmatrix} \begin{bmatrix} c_1 \\ c_2 \end{bmatrix} = \begin{bmatrix} 1 \\ -1 \end{bmatrix},$$

an equation in the form $V\mathbf{c} = \mathbf{b}$, which is easily solved in MATLAB.

```
>> b=[1;-1];
>> c=V\b
c =
    1.5356
   -1.3907
```

Therefore, the solution of system (10.13) is

$$\mathbf{x}(t) = 1.5356 e^{-8.2170t} \begin{bmatrix} 0.9814 \\ 0.1921 \end{bmatrix} - 1.3907 e^{1.2170t} \begin{bmatrix} 0.3646 \\ 0.9312 \end{bmatrix}.$$

Example 6. *Find the general solution of the system*

$$\begin{aligned}
x_1' &= 15x_1 - 6x_2 - 18x_3 - 6x_4, \\
x_2' &= -4x_1 + 5x_2 + 8x_3 + 4x_4, \\
x_3' &= 12x_1 - 6x_2 - 15x_3 - 6x_4, \\
x_4' &= 4x_1 - 2x_2 - 8x_3 - x_4.
\end{aligned} \tag{10.17}$$

Set up the system as a matrix equation.

$$\begin{bmatrix} x_1 \\ x_2 \\ x_3 \\ x_4 \end{bmatrix}' = \begin{bmatrix} 15 & -6 & -18 & -6 \\ -4 & 5 & 8 & 4 \\ 12 & -6 & -15 & -6 \\ 4 & -2 & -8 & -1 \end{bmatrix} \begin{bmatrix} x_1 \\ x_2 \\ x_3 \\ x_4 \end{bmatrix} \tag{10.18}$$

Equation (10.18) now has the form $\mathbf{x}' = A\mathbf{x}$. Enter the matrix A in MATLAB's workspace and compute the eigenvalues.

```
>> A=[15 -6 -18 -6;-4 5 8 4;12 -6 -15 -6;4 -2 -8 -1];
>> E=eig(A)
E =
   -3.0000
    3.0000
    1.0000
    3.0000
```

What sets this matrix apart from previous matrices is the presence of a *repeated* eigenvalue. The repeated eigenvalue 3 is an eigenvalue of multiplicity two[2].

The eigenvectors associated with $\lambda_1 = -3$ and $\lambda_2 = 1$ are easily captured with MATLAB's null command.

```
>> v1=null(A-(-3)*eye(4),'r')
v1 =
     1
    -1
     1
     1
```

```
>> v2=null(A-1*eye(4),'r')
v2 =
     0
    -1
     0
     1
```

The three equations in Example 4 required three independent eigenvectors to form the general solution. The two equations in Example 5 required two independent eigenvectors to form the general solution. In this example we have four equations, so four independent eigenvectors are required. We have two independent eigenvectors, but only one eigenvalue remains. Let's cross our fingers for luck and use MATLAB's null command once more.

---

[2] Symbolic Toolbox users might find the following sequence of commands illuminating: p=poly(sym(A));factor(p). Does this demonstrate why 3 is an eigenvalue of multiplicity two?

```
>> null(A-3*eye(4),'r')
ans =
    0.5000   -1.0000
    1.0000        0
         0   -1.0000
         0    1.0000
```

Fortunately for us[3], the eigenspace associated with the eigenvalue $\lambda_3 = 3$ has two basis vectors, exactly the number needed to craft the general solution

$$\mathbf{x}(t) = c_1 e^{-3t} \begin{bmatrix} 1 \\ -1 \\ 1 \\ 1 \end{bmatrix} + c_2 e^{1t} \begin{bmatrix} 0 \\ -1 \\ 0 \\ 1 \end{bmatrix} + c_3 e^{3t} \begin{bmatrix} 1/2 \\ 1 \\ 0 \\ 0 \end{bmatrix} + c_4 e^{3t} \begin{bmatrix} -1 \\ 0 \\ -1 \\ 1 \end{bmatrix}.$$

Complex Eigenvalues

Two key facts aid in the solution of systems involving complex eigenvalues and eigenvectors. Foremost is Euler's famous identity.

$$e^{i\theta} = \cos\theta + i\sin\theta$$

Secondly, if you find a complex solution of a linear, homogeneous system with constant *real* coefficients, then both the real and imaginary parts of your complex solution will also be solutions of your system.

Example 7. *Solve the initial value problem*

$$\begin{aligned} x_1' &= -x_1 + 2x_3, \\ x_2' &= 2x_1 + 3x_2 - 6x_3, \\ x_3' &= -2x_1 - x_3, \end{aligned} \tag{10.19}$$

where $x_1(0) = 4$, $x_2(0) = -5$, and $x_3(0) = 9$.

Set up the system as a matrix equation.

$$\begin{bmatrix} x_1 \\ x_2 \\ x_3 \end{bmatrix}' = \begin{bmatrix} -1 & 0 & 2 \\ 2 & 3 & -6 \\ -2 & 0 & -1 \end{bmatrix} \begin{bmatrix} x_1 \\ x_2 \\ x_3 \end{bmatrix} \tag{10.20}$$

Equation (10.20) has the form $\mathbf{x}' = A\mathbf{x}$. Note that the entries of the matrix A are real numbers, so the coefficients of the characteristic polynomial must be real numbers, which is easily checked in MATLAB.

```
>> A=[-1 0 2;2 3 -6;-2 0 -1];
>> p=poly(A)
p =
     1    -1    -1   -15
```

---

[3] In later sections of this chapter we will show how to form the general solution when you have fewer than the required number of eigenvectors.

174

If a polynomial has real coefficients, any complex zeros must occur in complex conjugate[4] pairs.

```
>> r=roots(p)
r =
    3.0000
   -1.0000 + 2.0000i
   -1.0000 - 2.0000i
```

Note that both $-1+2i$ and $-1-2i$ are eigenvalues; the eigenvalues occur in conjugate pairs. Furthermore, the eigenvectors must also occur in complex conjugate pairs[5].

```
>> [V,E]=eig(A)
V =
         0            0.5000            0.5000
    1.0000       -0.5000 + 0.5000i   -0.5000 - 0.5000i
         0            0 + 0.5000i         0 - 0.5000i
E =
    3.0000            0                 0
         0       -1.0000 + 2.0000i      0
         0            0             -1.0000 - 2.0000i
```

Note that the eigenvector in column two of matrix V is the complex conjugate of the eigenvector in column three of matrix V.

We can form the general solution if we can find three linearly independent solutions. First of all, since 3 is an eigenvalue with associated eigenvector $[0, 1, 0]^T$,

$$\mathbf{x}_1(t) = e^{3t} \begin{bmatrix} 0 \\ 1 \\ 0 \end{bmatrix}$$

is a solution of system (10.19).

The eigenvalue $-1 + 2i$ appears in $E(2,2)$ and we select a multiple of the second column of V, $[1, -1+i, i]$, as the associated eigenvector. Consequently, a complex solution of system (10.19) is

$$\mathbf{x}_c = e^{(-1+2i)t} \begin{bmatrix} 1 \\ -1+i \\ i \end{bmatrix} = e^{-t}e^{i2t} \left(\begin{bmatrix} 1 \\ -1 \\ 0 \end{bmatrix} + i \begin{bmatrix} 0 \\ 1 \\ 1 \end{bmatrix} \right).$$

Euler's formula and a little complex arithmetic reveals the following.

$$\mathbf{x}_c = e^{-t}(\cos 2t + i \sin 2t) \left(\begin{bmatrix} 1 \\ -1 \\ 0 \end{bmatrix} + i \begin{bmatrix} 0 \\ 1 \\ 1 \end{bmatrix} \right)$$

$$\mathbf{x}_c = \left(e^{-t} \begin{bmatrix} \cos 2t \\ -\cos 2t - \sin 2t \\ -\sin 2t \end{bmatrix} + ie^{-t} \begin{bmatrix} \sin 2t \\ -\sin 2t + \cos 2t \\ \cos 2t \end{bmatrix} \right)$$

---

[4] The conjugate of $a + bi$ is $a - bi$.

[5] The conjugate of a vector or matrix is found by taking the complex conjugate of each entry. For example, the conjugate of $[1/2, -1/2 + (1/2)i, (1/2)i]^T$ is $[1/2, -1/2 - (1/2)i, -(1/2)i]^T$.

Now we use the fact that the real and imaginary parts of x_c form real, independent solutions of system (10.19).

$$x_2(t) = e^{-t} \begin{bmatrix} \cos 2t \\ -\cos 2t - \sin 2t \\ -\sin 2t \end{bmatrix} \quad \text{and} \quad x_3(t) = e^{-t} \begin{bmatrix} \sin 2t \\ -\sin 2t + \cos 2t \\ \cos 2t \end{bmatrix}$$

We now have three independent solutions $x_1(t)$, $x_2(t)$, and $x_3(t)$, so we can form the general solution with $x(t) = c_1 x_1(t) + c_2 x_2(t) + c_3 x_3(t)$.

$$x(t) = c_1 e^{3t} \begin{bmatrix} 0 \\ 1 \\ 0 \end{bmatrix} + c_2 e^{-t} \begin{bmatrix} \cos 2t \\ -\cos 2t - \sin 2t \\ -\sin 2t \end{bmatrix} + c_3 e^{-t} \begin{bmatrix} \sin 2t \\ -\sin 2t + \cos 2t \\ \cos 2t \end{bmatrix}. \tag{10.21}$$

We need to determine the solution that satisfies the initial condition $x(0) = [x_1(0), x_2(0), x_3(0)]^T = [4, -5, 9]^T$. Substitute $t = 0$ in equation (10.21) to get

$$\begin{bmatrix} 4 \\ -5 \\ 9 \end{bmatrix} = c_1 \begin{bmatrix} 0 \\ 1 \\ 0 \end{bmatrix} + c_2 \begin{bmatrix} 1 \\ -1 \\ 0 \end{bmatrix} + c_3 \begin{bmatrix} 0 \\ 1 \\ 1 \end{bmatrix},$$

$$\begin{bmatrix} 0 & 1 & 0 \\ 1 & -1 & 1 \\ 0 & 0 & 1 \end{bmatrix} \begin{bmatrix} c_1 \\ c_2 \\ c_3 \end{bmatrix} = \begin{bmatrix} 4 \\ -5 \\ 9 \end{bmatrix},$$

a matrix equation in the form $Wc = b$.

```
>> W=[0 1 0;1 -1 1;0 0 1];
>> b=[4;-5;9];
>> c=W\b
c =
    -10
      4
      9
```

Substitution of $c_1 = -10$, $c_2 = 4$, and $c_3 = 9$ in equation (10.21), a little arithmetic, and we arrive at the final solution of the initial value problem posed in equation (10.19).

$$\begin{bmatrix} x_1(t) \\ x_2(t) \\ x_3(t) \end{bmatrix} = \begin{bmatrix} e^{-t}(4\cos 2t + 9\sin 2t) \\ -10e^{3t} + e^{-t}(5\cos 2t - 13\sin 2t) \\ e^{-t}(9\cos 2t - 4\sin 2t) \end{bmatrix}$$

The Exponential Matrix

We take pause from solving systems of differential equations to explore a very important matrix operator, the exponential of a matrix.

Definition. *If A is a $n \times n$ matrix, then the exponential of A, symbolized by e^A, is defined as*

$$e^A = I + A + \frac{A^2}{2!} + \frac{A^3}{3!} + \cdots + \frac{A^k}{k!} + \cdots, \tag{10.22}$$

where I is the $n \times n$ identity matrix.

176

It can be shown that the infinite series (10.22) converges for every $n \times n$ matrix A. However, in all but a few simple cases, it's usually quite difficult to compute the exponential of a matrix with this definition.

Diagonal matrices are always easy to work with. If

$$D = \begin{bmatrix} \lambda_1 & 0 & \cdots & 0 \\ 0 & \lambda_2 & \cdots & 0 \\ \vdots & \vdots & \ddots & \vdots \\ 0 & 0 & \cdots & \lambda_n \end{bmatrix},$$

then

$$e^D = I + D + \frac{1}{2!}D^2 + \cdots,$$

$$= \begin{bmatrix} 1 + \lambda_1 + \frac{1}{2!}\lambda_1^2 + \cdots & 0 & \cdots & 0 \\ 0 & 1 + \lambda_2 + \frac{1}{2!}\lambda_2^2 + \cdots & \cdots & 0 \\ \vdots & \vdots & \ddots & \vdots \\ 0 & 0 & \cdots & 1 + \lambda_n + \frac{1}{2!}\lambda_n^2 + \cdots \end{bmatrix},$$

$$= \begin{bmatrix} e^{\lambda_1} & 0 & \cdots & 0 \\ 0 & e^{\lambda_2} & \cdots & 0 \\ \vdots & \vdots & \ddots & \vdots \\ 0 & 0 & \cdots & e^{\lambda_n} \end{bmatrix}.$$

Example 8. *If I is the $n \times n$ identity matrix and r is a scalar (number), prove that $e^{rI} = e^r I$.*

This result is immediate if we use the previously developed fact about diagonal matrices.

$$e^{rI} = e^{\begin{bmatrix} r & \cdots & 0 \\ \vdots & \ddots & \vdots \\ 0 & \cdots & r \end{bmatrix}} = \begin{bmatrix} e^r & \cdots & 0 \\ \vdots & \ddots & \vdots \\ 0 & \cdots & e^r \end{bmatrix} = e^r \begin{bmatrix} 1 & \cdots & 0 \\ \vdots & \ddots & \vdots \\ 0 & \cdots & 1 \end{bmatrix} = e^r I$$

For example,

$$e^{3I} = e^3 I = e^3 \begin{bmatrix} 1 & 0 \\ 0 & 1 \end{bmatrix} = \begin{bmatrix} e^3 & 0 \\ 0 & e^3 \end{bmatrix}.$$

You can easily check this result with MATLAB's expm command[6]. Note that $e^3 \approx 20.0855$.

```
>> I=eye(2);
>> E=expm(3*I)
E =
    20.0855         0
         0    20.0855
```

Proposition 1. *If the matrices A and B commute, i.e., if $AB = BA$, then*

$$e^{A+B} = e^A e^B.$$

---

[6] It is important to note that exp(3*eye(2)) gives an incorrect result. Try it! You must always use expm when computing e^A, the exponential of matrix A. Symbolic Toolbox users will want to try expm(3*sym(eye(2))).

For example, note that A=[1,2;-1,0] and B=[0,-2;1,1] commute (enter A*B and B*A and compare). Compute

```
>> expm(A+B)
ans =
     2.7183         0
          0    2.7183

>> expm(A)*expm(B)
ans =
     2.7183         0
          0    2.7183
```

and note that $e^{A+B} = e^A e^B$ for these particular matrices A and B. However, matrices C=[1,1;1,1] and D=[1,2;3,4] do not commute (Check this with C*D and D*C.). Use MATLAB to show that $e^{C+D} \neq e^C e^D$.

Proposition 2. *If A is an $n \times n$ matrix, then the derivative of e^{tA} is given as*

$$\frac{d}{dt}e^{tA} = Ae^{tA}.$$

Symbolic Toolbox Users. Owners of the Symbolic Toolbox can readily check the validity of this last proposition for the matrix A=[1,-1;-1,1].

```
>> syms t
>> A=sym([1,-1;-1,1]);
>> diff(expm(t*A),t)
ans =
[  exp(2*t),  -exp(2*t)]
[ -exp(2*t),   exp(2*t)]
>> A*expm(t*A)
ans =
[  exp(2*t),  -exp(2*t)]
[ -exp(2*t),   exp(2*t)]
```

Note that the results are identical, showing that $\frac{d}{dt}e^{tA} = Ae^{tA}$ for this particular matrix A. Experiment! Convince yourself that $\frac{d}{dt}e^{tA} = Ae^{tA}$ for other square matrices A. Try a 3×3 matrix.

We end this section with a powerful theorem. One of the first initial value problems encountered in a course on ordinary differential equations is $x' = ax$, $x(0) = x_0$. It is easy to show (separate the variables) that the solution of this initial value problem is $x = e^{at}x_0$. One would hope that a similar result applies to the matrix equation $\mathbf{x}' = A\mathbf{x}$, $\mathbf{x}(0) = \mathbf{x}_0$.

Theorem 1. *Let $\mathbf{x}' = A\mathbf{x}$ be the matrix equation representing a system of n first order, linear, homogeneous differential equations with constant coefficients. The solution of the initial value problem, $\mathbf{x}' = A\mathbf{x}$, $\mathbf{x}(0) = \mathbf{x}_0$, is*

$$\mathbf{x}(t) = e^{At}\mathbf{x}_0.$$

Readers can easily prove this theorem by direct substitution and the application of Proposition 2. The initial condition is easily verified using the result developed in Example 8.

Symbolic Toolbox Users. In Example 4, we used eigenvalues and eigenvectors to show that the solution of the system

$$\begin{bmatrix} x_1 \\ x_2 \\ x_3 \end{bmatrix}' = \begin{bmatrix} 8 & -5 & 10 \\ 2 & 1 & 2 \\ -4 & 4 & -6 \end{bmatrix} \begin{bmatrix} x_1 \\ x_2 \\ x_3 \end{bmatrix}, \tag{10.10}$$

with initial condition $\mathbf{x}_0 = [2, 2, -3]^T$, was

$$\begin{bmatrix} x_1(t) \\ x_2(t) \\ x_3(t) \end{bmatrix} = \begin{bmatrix} 6e^{-2t} - 4e^{3t} \\ -4e^{3t} + 6e^{2t} \\ -6e^{-2t} + 3e^{2t} \end{bmatrix}.$$

Let's use Theorem 1 to produce the same result.

```
>> syms t
>> A=[8 -5 10;2 1 2;-4 4 -6];
>> x0=sym([2;2;-3]);
>> x=expm(A*t)*x0
x =
[ 6*exp(-2*t)-4*exp(3*t)]
[ 6*exp(2*t)-4*exp(3*t)]
[ 3*exp(2*t)-6*exp(-2*t)]
```

This is a startling result, clearly demonstrating the power of the exponential matrix.

Repeated Eigenvalues

When the eigenvalues of matrix A are distinct, then the matrix is *diagonalizable* (See Exercises 16–19) and finding e^{At} is not too difficult, but the presence of repeated eigenvalues complicates the process. The matrix A in Example 6 had a repeated eigenvalue, but it still had a full complement of eigenvectors; and that allowed us to complete the solution. However, if the matrix A is $n \times n$ and there are fewer than n independent eigenvectors, then the level of difficulty is raised an order or two in magnitude.

Example 9. *Find the solution of the initial value problem*

$$\begin{aligned} x_1' &= 13x_1 + 11x_2, \\ x_2' &= -11x_1 - 9x_2, \end{aligned} \tag{10.23}$$

with initial conditions $x_1(0) = 1$ and $x_2(0) = 2$.

System (10.23) has matrix form $\mathbf{x}' = A\mathbf{x}$, as in

$$\begin{bmatrix} x_1 \\ x_2 \end{bmatrix}' = \begin{bmatrix} 13 & 11 \\ -11 & -9 \end{bmatrix} \begin{bmatrix} x_1 \\ x_2 \end{bmatrix}, \tag{10.24}$$

with initial condition $\mathbf{x}(0) = [1, 2]^T$. It is easily verified (try poly(A)) that the characteristic polynomial is $p(\lambda) = \lambda^2 - 4\lambda + 4$, which factors $p(\lambda) = (\lambda - 2)^2$, indicating an eigenvalue $\lambda = 2$ of algebraic multiplicity two. Difficulties become immediately apparent when we execute the command

179

```
>> null(A-2*eye(2),'r')
ans =

    -1
     1
```

showing that, up to multiplication by a constant, $v_1 = [-1, 1]^T$ is the only eigenvector. We need two independent solutions to form the general solution of system (10.23), but we only have one.

We need some terminology to describe this phenomenon.

Definition. *If λ_j is an eigenvalue of the matrix A, we define the **algebraic multiplicity** of λ_j to be the largest power of $(\lambda - \lambda_j)$ which divides the characteristic polynomial of A. We define the **geometric multiplicity** of λ_j to be the dimension of the nullspace of $A - \lambda_j I$, i.e. the dimension of the **eigenspace** associated to λ_j.*

It can be shown that the geometric multiplicity is always smaller than or equal to the algebraic multiplicity. In the case of system (10.24), the geometric multiplicity (one) of the eigenvalue $\lambda = 2$ is strictly smaller than its algebraic multiplicity (two).

However, in the case of system (10.24), $tA = 2tI + t(A - 2I)$. Furthermore, since the identity matrix commutes with every other matrix, the matrices $2tI$ and $t(A - 2I)$ commute. Hence, we can use Proposition 1 and the identity developed in Example 8 to write

$$
\begin{aligned}
e^{tA} &= e^{2tI+t(A-2I)}, \\
&= e^{2tI} e^{t(A-2I)}, \\
&= e^{2t} I e^{t(A-2I)}, \\
&= e^{2t} e^{t(A-2I)}.
\end{aligned}
\tag{10.25}
$$

We can use the definition of the exponential of a matrix to create a series expansion of the second exponential in equation (10.25).

$$
e^{tA} = e^{2t} \left(I + t(A - 2I) + \frac{t^2}{2!}(A - 2I)^2 + \frac{t^3}{3!}(A - 2I)^3 + \cdots \right)
\tag{10.26}
$$

However, because 2 is the *only* eigenvalue of the 2×2 matrix A (See Exercises 20–21),

```
>> (A-2*eye(2))^2
ans =
     0     0
     0     0
```

Therefore, $(A - 2I)^k = 0$ for all $k \geq 2$ and

$$
\begin{aligned}
e^{tA} &= e^{2t} \left(I + t(A - 2I) \right), \\
&= e^{2t} \left\{ \begin{bmatrix} 1 & 0 \\ 0 & 1 \end{bmatrix} + t \left(\begin{bmatrix} 13 & 11 \\ -11 & -9 \end{bmatrix} - 2 \begin{bmatrix} 1 & 0 \\ 0 & 1 \end{bmatrix} \right) \right\}, \\
&= e^{2t} \begin{bmatrix} 1+11t & 11t \\ -11t & 1-11t \end{bmatrix}.
\end{aligned}
$$

180

According to Theorem 1, the solution of system (10.24), with initial condition $\mathbf{x}(0) = [1, 2]^T$, is given by $\mathbf{x}(t) = e^{tA}\mathbf{x}(0)$. Thus,

$$\mathbf{x}(t) = e^{tA}\mathbf{x}(0),$$
$$= e^{2t}\left(\begin{bmatrix} 1+11t & 11t \\ -11t & 1-11t \end{bmatrix}\right)\begin{bmatrix} 1 \\ 2 \end{bmatrix},$$
$$= e^{2t}\begin{bmatrix} 1+33t \\ 2-33t \end{bmatrix}.$$

Symbolic Toolbox Users. Owners of the Symbolic Toolbox can easily check this last solution.

```
>> syms t
>> A=sym([13 11;-11 -9]);
>> x0=sym([1;2]);
>> x=expm(t*A)*x0
x =
[     exp(2*t)+33*t*exp(2*t)]
[ -33*t*exp(2*t)+2*exp(2*t)]
```

In the case of system (10.24) we were able to actually compute the exponential of the matrix tA. It is rare that this is so easy, but the method we used works for any 2 by 2 matrix which has a single eigenvalue λ of multiplicity 2. In such a case we always have

$$e^{tA} = e^{\lambda t}\left(I + t(A - \lambda I)\right),$$

and the solution to the initial value problem is given by

$$\mathbf{x}(t) = e^{tA}\mathbf{x}_0 = e^{\lambda t}\left(I + t(A - \lambda I)\right)\mathbf{x}_0.$$

Let's do a more complicated example.

Example 10. *Solve the initial value problem*

$$\begin{aligned} x_1' &= 3x_1 - 3x_2 - 6x_3 + 5x_4, \\ x_2' &= -3x_1 + 2x_2 + 5x_3 - 4x_4, \\ x_3' &= 2x_1 - 6x_2 - 4x_3 + 7x_4, \\ x_4' &= -3x_1 + 5x_3 - 2x_4, \end{aligned} \tag{10.27}$$

where $x_1(0) = 2$, $x_2(0) = -1$, $x_3(0) = 0$, and $x_4(0) = 0$.

What made our solution to the previous case possible was that $(A - 2I)^2 = 0$. From this we were able to conclude that the infinite series in (10.26) was actually a finite series which we were able to compute. We cannot expect to be so lucky this time.

Our system can be written

$$\mathbf{x}' = A\mathbf{x},$$

where

$$A = \begin{bmatrix} 3 & -3 & -6 & 5 \\ -3 & 2 & 5 & -4 \\ 2 & -6 & -4 & 7 \\ -3 & 0 & 5 & -2 \end{bmatrix}.$$

Let's use `eig` to compute the eigenvalues.

```
>> A=[3 -3 -6 5;-3 2 5 -4;2 -6 -4 7;-3 0 5 -2];
>> E = eig(A)
E =
  -1.0000 + 0.0000i
  -1.0000 - 0.0000i
  -1.0000
   2.0000
```

We suspect roundoff error and conjecture that -1 is an eigenvalue of multiplicity 3. If the Symbolic Toolbox is available, this is confirmed by the command

```
>> eig(sym(A))
ans =
[ 2]
[-1]
[-1]
[-1]
```

Thus, we see that A has two different eigenvalues, -1 and 2, with multiplicities 3 and 1, respectively[7]. The fact that there are two different eigenvalues means that we will not be able to find any number λ for which $(A - \lambda I)^k = 0$, for large values of k, which is necessary to make the series

$$e^{tA} = e^{\lambda t I} e^{t(A - \lambda I)} = e^{\lambda t} \left(I + t(A - \lambda I) + \frac{t^2}{2!}(A - \lambda I)^2 + \frac{t^3}{3!}(A - \lambda I)^3 + \cdots \right) \qquad (10.28)$$

reduce to a finite series. However, it is not really necessary that (10.28) truncate in this manner. Let's apply (10.28) to a vector **w**.

$$e^{tA}\mathbf{w} = e^{\lambda t} \left(I\mathbf{w} + t(A - \lambda I)\mathbf{w} + \frac{t^2}{2!}(A - \lambda I)^2\mathbf{w} + \frac{t^3}{3!}(A - \lambda I)^3\mathbf{w} + \cdots \right) \qquad (10.29)$$

For this series to truncate, it is only necessary that $(A - \lambda I)^k\mathbf{w} = 0$ for all large k, *for this particular vector*. Let's give this kind of vector a name.

Definition. *Suppose λ is an eigenvalue of the matrix A. Any vector* **w** *which satisfies* $(A - \lambda I)^k \mathbf{w} = 0$ *for some integer k is called a generalized eigenvector associated with the eigenvalue λ.*

---

[7] Owners of the Symbolic Toolbox can further confirm the algebraic multiplicity of the eigenvalues with `ps=poly(sym(A)),factor(ps)`.

Is this really a step forward? The fact is that there are always enough generalized eigenvectors to find a basis for solutions. We will state the actual result as a theorem.

Theorem 2. *Suppose A is a matrix and λ is an eigenvalue of algebraic multiplicity d. Then there is an integer $k \leq d$ such that the nullspace of $(A - \lambda I)^k$ is the set of all generalized eigenvectors associated to λ. The nullspace will have dimension equal to d, the algebraic multiplicity of λ.*

In Example 9, we found that $(A - 2I)^2 = 0$. This meant that every two vector was a generalized eigenvector associated to the eigenvalue 2. It was the fact that every vector was a generalized eigenvector that made things so easy in that case.

In order to complete Example 10 we will have to work a little harder. Let's look at what happens for the eigenvalue -1, which has algebraic multiplicity 3. We will use null to compute the generalized eigenvectors, which are the vectors in the nullspaces of powers of $A - (-1)I = A + I$. First it will be convenient to introduce some shorthand.

```
>> I=eye(4);
```

Then we get

```
>> null(A+I,'r')
ans =
    -2
     1
    -1
     1
```

We see that although -1 is an eigenvalue of algebraic multiplicity 3, the eigenspace has dimension 1; i.e., the eigenvalue -1 has geometric multiplicity 1.

Now let's look for generalized eigenvectors by examining the nullspace of higher powers of $A + I$. First,

```
>> null((A+I)^2,'r')
ans =
     2     0
     0     1
     1     0
     0     1
```

shows that the dimension has increased to 2. However, since the algebraic multiplicity of the eigenvalue is 3, we want more. We proceed to

```
>> null((A+I)^3,'r')
ans =
     1     0     0
     0     0     1
     0     1     0
     0     0     1
```

Now the dimension is 3, the same as the algebraic multiplicity, so we are in business. For each vector in the nullspace of $(A+I)^3$ the series in (10.30) will become finite and we will be able to compute the solution.

We want to compute the series for each vector in a basis of the nullspace of $(A+I)^3$. We could choose the three column vectors found by `null`, but there is a systematic way to proceed that will reduce the algebra somewhat. Here is the plan:

- Choose a vector $\mathbf{v}_1$ which is in the nullspace of $(A+I)^3$, but **not** in the nullspace of $(A+I)^2$. That is, $\mathbf{v}_1$ must not be a linear combination of the basis vectors found for the nullspace of $(A+I)^2$. Of course, this makes $(A+I)^3\mathbf{v}_1 = \mathbf{0}$.
- Set $\mathbf{v}_2 = (A+I)\mathbf{v}_1$. Then $(A+I)^2\mathbf{v}_2 = (A+I)^3\mathbf{v}_1 = \mathbf{0}$.
- Set $\mathbf{v}_3 = (A+I)\mathbf{v}_2$. Then $\mathbf{v}_3 = (A+I)^2\mathbf{v}_1$, and $(A+I)\mathbf{v}_3 = (A+I)^2\mathbf{v}_2 = (A+I)^3\mathbf{v}_1 = \mathbf{0}$.

The result is that the vectors are automatically linearly independent, and because of the algebraic relations between them the computations become a lot easier.

There are a lot of possibilities in the choice of $\mathbf{v}_1$, but we choose

```
>>  v1 = [1 0 0 0]'
v1 =
        1
        0
        0
        0
```

Then we set

```
>> v2 = (A+I)*v1
v2 =
        4
       -3
        2
       -3
```

and finally set

```
>> v3 = (A+I)*v2
v3 =
       -2
        1
       -1
        1
```

Now we can use (10.29) to compute the solution to the initial value problem with each of these initial conditions. The first computation is the hardest, but notice how knowing the algebraic relations of the basis vectors speeds things up. Since $(A + I)^3\mathbf{v}_1 = \mathbf{0}$, we have that $(A + I)^k\mathbf{v}_1 = \mathbf{0}$ for all $k \geq 3$.

$$\mathbf{x}_1(t) = e^{tA}\,\mathbf{v}_1$$

$$= e^{-t}\left(\mathbf{v}_1 + t(A + I)\,\mathbf{v}_1 + \frac{t^2}{2!}(A + I)^2\mathbf{v}_1\right)$$

$$= e^{-t}\left(\mathbf{v}_1 + t\mathbf{v}_2 + \frac{t^2}{2!}\mathbf{v}_3\right)$$

$$= e^{-t}\left(\begin{bmatrix} 1 \\ 0 \\ 0 \\ 0 \end{bmatrix} + t\begin{bmatrix} 4 \\ -3 \\ 2 \\ -3 \end{bmatrix} + \frac{t^2}{2!}\begin{bmatrix} -2 \\ 1 \\ -1 \\ 1 \end{bmatrix}\right)$$

$$= e^{-t}\begin{bmatrix} 1 + 4t - t^2 \\ -3t + t^2/2 \\ 2t - t^2/2 \\ -3t + t^2/2 \end{bmatrix}.$$

The other computations are shorter. Since $(A + I)^2\mathbf{v}_2 = \mathbf{0}$, $(A + I)^k\mathbf{v}_2 = \mathbf{0}$ for all $k \geq 2$. We have

$$\mathbf{x}_2(t) = e^{tA}\,\mathbf{v}_2,$$

$$= e^{-t}\left(\mathbf{v}_2 + t(A + I)\,\mathbf{v}_2\right),$$

$$= e^{-t}\left(\mathbf{v}_2 + t\mathbf{v}_3\right),$$

$$= e^{-t}\left(\begin{bmatrix} 4 \\ -3 \\ 2 \\ -3 \end{bmatrix} + t\begin{bmatrix} -2 \\ 1 \\ -1 \\ 1 \end{bmatrix}\right),$$

$$= e^{-t}\begin{bmatrix} 4 - 2t \\ -3 + t \\ 2 - t \\ -3 + t \end{bmatrix}.$$

and , since $(A + I)\mathbf{v}_3 = \mathbf{0}$, $(A + I)^k\mathbf{v}_3 = \mathbf{0}$ for all $k \geq 1$,

$$\mathbf{x}_3(t) = e^{tA}\,\mathbf{v}_3,$$

$$= e^{-t}\,\mathbf{v}_3,$$

$$= e^{-t}\begin{bmatrix} -2 \\ 1 \\ -1 \\ 1 \end{bmatrix}.$$

Thus, we have found three linearly independent solutions corresponding to the eigenvalue -1, which has algebraic multiplicity three. Since A is a 4 by 4 matrix we need one more. This can be found in the usual way from the eigenvalue 2.

```
v4=null(A-2*I,'r')
v4 =
   -0.5000
    0.5000
    0.5000
    1.0000
```

The solution to the initial value problem with this initial condition is

$$\mathbf{x}_4(t) = e^{tA}\mathbf{v}_4,$$
$$= e^{2t}\mathbf{v}_4,$$
$$= e^{2t}\begin{bmatrix} -1/2 \\ 1/2 \\ 1/2 \\ 1 \end{bmatrix}.$$

Having found four linearly independent solutions, linearity allows us to form the general solution of system (10.27) by forming a linear combination of our basic solutions.

$$\mathbf{x}(t) = c_1\mathbf{x}_1(t) + c_2\mathbf{x}_2(t) + c_3\mathbf{x}_3(t) + c_4\mathbf{x}_4(t) \tag{10.30}$$

To solve the initial value problem with $\mathbf{x}(0) = \mathbf{x}_0 = [2, -1, 0, 0]^T$, we must substitute $t = 0$ in equation (10.30).

$$\mathbf{x}(0) = c_1\mathbf{x}_1(0) + c_2\mathbf{x}_2(0) + c_3\mathbf{x}_3(0) + c_4\mathbf{x}_4(0)$$
$$\mathbf{x}_0 = c_1\mathbf{v}_1 + c_2\mathbf{v}_2 + c_3\mathbf{v}_3 + c_4\mathbf{v}_4$$

Of course, this is equivalent to the matrix equation

$$\mathbf{x}_0 = [\mathbf{v}_1, \mathbf{v}_2, \mathbf{v}_3, \mathbf{v}_4]\begin{bmatrix} c_1 \\ c_2 \\ c_3 \\ c_4 \end{bmatrix},$$

so we can find the solution with the following MATLAB code.

```
>> V = [v1 v2 v3 v4];
>> x0 = [2;-1;0;0];
>> c=V\x0
c =
    5.0000
    3.0000
    7.0000
    2.0000
```

Therefore, the solution of system (10.27) is $\mathbf{x}(t) = 5\mathbf{x}_1(t) + 3\mathbf{x}_2(t) + 7\mathbf{x}_3(t) + 2\mathbf{x}_4(t)$, or

$$\mathbf{x}(t) = 5e^{-t}\begin{bmatrix} 1 + 4t - t^2 \\ -3t + t^2/2 \\ 2t - t^2/2 \\ -3t + t^2/2 \end{bmatrix} + 3e^{-t}\begin{bmatrix} 4 - 2t \\ -3 + t \\ 2 - t \\ -3 + t \end{bmatrix} + 7e^{-t}\begin{bmatrix} -2 \\ 1 \\ -1 \\ 1 \end{bmatrix} + 2e^{2t}\begin{bmatrix} -1/2 \\ 1/2 \\ 1/2 \\ 1 \end{bmatrix}.$$

Users of the Symbolic Toolbox might want to compare this last solution with

```
>> syms t
>> A=sym(A);
>> x0=sym(x0);
>> x=expm(A*t)*x0
x =
[      3*exp(-t)+14*t*exp(-t)-5*t^2*exp(-t)-exp(2*t)]
[  -12*t*exp(-t)+5/2*t^2*exp(-t)+exp(2*t)-2*exp(-t)]
[      7*t*exp(-t)-5/2*t^2*exp(-t)-exp(-t)+exp(2*t)]
[ -12*t*exp(-t)+5/2*t^2*exp(-t)-2*exp(-t)+2*exp(2*t)]
```

Exercises

1. Consider the following matrices:

   a) $\begin{bmatrix} -8 & 10 \\ -5 & 7 \end{bmatrix}$
   b) $\begin{bmatrix} -4 & 6 & 0 \\ -3 & 5 & 0 \\ -2 & 0 & 1 \end{bmatrix}$
   c) $\begin{bmatrix} -5 & -18 & 12 \\ -3 & -8 & 6 \\ -6 & -18 & 13 \end{bmatrix}$
   d) $\begin{bmatrix} 2 & -3 & -1 & -3 \\ 3 & -4 & 3 & -3 \\ 0 & 0 & 3 & 0 \\ -3 & 3 & -3 & 2 \end{bmatrix}$

   For each matrix, perform each of the following tasks.

   i) Use the command p=poly(A) to find the characteristic polynomial.

   ii) Plot the characteristic polynomial p with the commands eig

   ```
   >> t=linspace(-4,4);
   >> y=polyval(p,t);
   >> plot(t,y)
   >> grid on
   >> axis([-4,4,-10,10])
   ```

   and find the roots of the polynomial with r=roots(p).

   iii) Find the eigenvalues with e=eig(A) and compare the eigenvalues with the roots of the polynomial found in part (ii). Where are the eigenvalues located on the graph produced in part (ii)?

2. **Symbolic Toolbox Users.** Enter each matrix from Exercise 1 as a symbolic object, as in A=sym([-8,10;-5,7]). Find the characteristic polynomial with p=poly(A), plot with ezplot(p), set the window and grid with axis([-4,4,-10 10]), grid on, factor with factor(p), and find the roots with solve(p). Finally, find the eigenvalues with eig(A) and compare with the roots of the characteristic polynomial.

3. Given that $\lambda = -1$ is an eigenvalue of the matrix

   $$A = \begin{bmatrix} 3 & 6 & 10 \\ 0 & -1 & 0 \\ 0 & -1 & -2 \end{bmatrix},$$

   reduce the augmented matrix M=[A-(-1)*eye(3),zeros(3,1)] and interpret the result to find the associated eigenvector. Use null(A-(-1)*eye(3),'r') and compare results. Perform similar analysis for the remaining eigenvalues, $\lambda = -2$ and $\lambda = 3$.

4. Consider again matrix A from Exercise 3. Use [V,E]=eig(A) to find the eigenvalues and eigenvectors of the matrix A. Strip off column three of matrix V with v3=V(:,3). Set v3=v3/v3(3). In a similar manner, show that the eigenvectors in the remaining columns of matrix V are multiples of the eigenvectors found in Exercise 3.

5. **Symbolic Toolbox Users.** Use [V,E]=eig(sym(A)) to find the eigenvalues and eigenvectors of matrix A from Exercise 3. Compare with the results found in Exercise 3.

6. Given that $\lambda = 1$ is an eigenvalue of matrix

   $$A = \begin{bmatrix} 5 & 5 & -12 & 18 \\ 4 & 6 & -12 & 18 \\ 0 & 6 & -11 & 12 \\ -2 & 2 & -3 & 1 \end{bmatrix},$$

   reduce the augmented matrix M=[A-1*eye(4),zeros(4,1)] and interpret the result to find a basis for the associated eigenspace. Use null(A-1*eye(4),'r') and compare the results. Perform similar analysis for the remaining eigenvalue, $\lambda = -2$. Clearly state the algebraic and geometric multiplicity of each eigenvalue.

7. For each of the following matrices find the eigenvalues and eigenvectors. You may use any method you wish as long as you choose eigenvectors with all integer entries. For each eigenvalue state the algebraic and the geometric multiplicity.

a) $\begin{bmatrix} -6 & 0 \\ 7 & 1 \end{bmatrix}$
b) $\begin{bmatrix} 0 & 1 \\ 1 & 0 \end{bmatrix}$
c) $\begin{bmatrix} 4 & 1 \\ -4 & 0 \end{bmatrix}$
d) $\begin{bmatrix} 2 & -1 \\ 1 & 0 \end{bmatrix}$

e) $\begin{bmatrix} -2 & 2 & -1 \\ -6 & 5 & -2 \\ 4 & -2 & 3 \end{bmatrix}$
f) $\begin{bmatrix} -3 & -2 & 3 \\ -2 & 1 & 1 \\ -7 & -3 & 6 \end{bmatrix}$
g) $\begin{bmatrix} -1 & 4 & -10 & 2 \\ 2 & -1 & 7 & -1 \\ 2 & -3 & 9 & -1 \\ 0 & -2 & 2 & 3 \end{bmatrix}$

8. **Real, Distinct Eigenvalues.** If the eigenvalues of matrix A are real and distinct, then the eigenvectors will be independent, providing the appropriate number of independent solutions for $\mathbf{x}' = A\mathbf{x}$ (See Example 4). Form the general solution of each of the following systems. Please choose eigenvectors with integer entries.

a) $\begin{aligned} x' &= -10x + 7y \\ y' &= -14x + 11y \end{aligned}$
b) $\begin{aligned} x' &= -15x + 18y - 8z \\ y' &= -4x + 4y - 2z \\ z' &= 16x - 21y + 9z \end{aligned}$
c) $\begin{aligned} x_1' &= 12x_1 - 16x_2 - 10x_3 - 22x_4 \\ x_2' &= 2x_1 - 4x_2 - 2x_3 - 4x_4 \\ x_3' &= -5x_1 + 9x_2 + 7x_3 + 10x_4 \\ x_4' &= 8x_1 - 11x_2 - 8x_3 - 15x_4 \end{aligned}$

9. **Symbolic Toolbox Users.** Check the solution of each system in Exercise 8 with the `dsolve` command (See Chapter 8). For example, the solution of Exercise 8a is obtained with the command
   `[x,y]=dsolve('Dx=-10*x+7*y','Dy=-14*x+11*y','t')`.

10. **Algebraic Multiplicity Equals Geometric Multiplicity.** If the algebraic multiplicity of each eigenvalue equals its geometric multiplicity, once again you will have sufficient eigenvectors to form the general solution (See Example 6). Form the general solution of each of the following systems. Please choose eigenvectors with integer entries.

a) $\begin{aligned} x' &= -5x - 3y - 3z \\ y' &= -6x - 2y - 3z \\ z' &= 12x + 6y + 7z \end{aligned}$
b) $\begin{aligned} w' &= 16w - 12x + 3y + 3z \\ x' &= 36w - 26x + 6y + 6z \\ y' &= 18w - 12x + y + 3z \\ z' &= 36w - 24x + 6y + 4z \end{aligned}$
c) $\begin{aligned} w' &= w + 3x + 3z \\ x' &= 3w + x + 3z \\ y' &= -6w + y - 6z \\ z' &= -3w - 3x - 5z \end{aligned}$

11. **Complex Eigenvalues.** The following systems have eigenvalues that are complex (See Example 7). Form the general solution of each system. Please choose eigenvectors whose real and imaginary parts contain integer entries.

a) $\begin{aligned} x' &= -7x + 15y \\ y' &= -6x + 11y \end{aligned}$
b) $\begin{aligned} x' &= 4x + 2y + z \\ y' &= -10x - 5y - 2z \\ z' &= 10x + 6y + 2z \end{aligned}$
c) $\begin{aligned} w' &= w + x - 2y - z \\ x' &= -5w - 2x + 6y + 3z \\ y' &= -5w - 2x + 5y + 2z \\ z' &= 10w + 5x - 10y - 4z \end{aligned}$

12. **Initial Value Problems.** If each of the exercises that follow, you are given matrix A and initial condition $\mathbf{x}_0$ of the system $\mathbf{x}' = A\mathbf{x}$, $\mathbf{x}(0) = \mathbf{x}_0$. Find a solution to the initial value problem. Again, eliminate fractions and decimals by using eigenvectors whose real and imaginary parts contain integer entries.

a) $\begin{bmatrix} 2 & 5 & -5 \\ -3 & -4 & 3 \\ -3 & -1 & 0 \end{bmatrix}, \begin{bmatrix} 1 \\ 1 \\ 1 \end{bmatrix}$
b) $\begin{bmatrix} -31 & 42 & 14 \\ -14 & 18 & 7 \\ -28 & 42 & 11 \end{bmatrix}, \begin{bmatrix} 1 \\ 2 \\ 1 \end{bmatrix}$
c) $\begin{bmatrix} -6 & 2 & 1 & 3 \\ -10 & 5 & 2 & 4 \\ 0 & 0 & 1 & 0 \\ -11 & 2 & 1 & 6 \end{bmatrix}, \begin{bmatrix} 3 \\ 2 \\ 1 \\ 1 \end{bmatrix}$

13. **Symbolic Toolbox Users.** Use the `dsolve` command to solve each of the initial value problems in Exercise 12. Compare these to the solutions found in Exercise 12. *Note: It is definitely challenging to get these solutions to match. Be persistent!*

14. **Numerical Versus Exact.** Provide time plots over the time interval $[0, 2\pi]$ for each of the solutions in Exercise 12. Place the plot of each dependent variable versus time on the same axes. Once that is completed, find a numerical solution for each system in Exercise 12 using `ode45` over the same time interval $[0, 2\pi]$. Provide a time plot of each dependent variable versus time on a second axes. How well do the exact and numerical solutions compare?

15. Use definition (10.22) of the exponential matrix to simplify each of the following.

188

a) e^A, where $A = \begin{bmatrix} 0 & 0 \\ 0 & 0 \end{bmatrix}$ b) e^{At}, where $A = \begin{bmatrix} 0 & 5 \\ 0 & 0 \end{bmatrix}$ c) e^{At}, where $A = \begin{bmatrix} 2 & 0 \\ 0 & 3 \end{bmatrix}$

If you own the Symbolic Toolbox, use it to check your results.

16. Create a diagonal matrix D with the command D=diag([1,2,3]). It is easy to calculate powers of a diagonal matrix. Enter D^2, D^3, D^4, etc. Do you see a pattern emerging? If

$$D = \begin{bmatrix} \lambda_1 & 0 & \cdots & 0 \\ 0 & \lambda_2 & \cdots & 0 \\ \vdots & \vdots & \ddots & \vdots \\ 0 & 0 & \cdots & \lambda_n \end{bmatrix},$$

then what is D^k?

17. If a matrix A has distinct eigenvalues (no repeated eigenvalues), then the matrix A can be *diagonalized*; i.e., there exists a diagonal matrix D and an invertible matrix V such that $A = VDV^{-1}$. Furthermore, if a matrix A is diagonalizable as $A = VDV^{-1}$, then the matrix D contains the eigenvalues of the matrix A on its diagonal and the columns of matrix V contain the eigenvectors of matrix A associated with the eigenvalues in the corresponding columns of matrix D. For each of the following matrices A, compute the eigenvalues and eigenvectors with [V,D]=eig(A), then show that $A = VDV^{-1}$ by comparing A with V*D*inv(V).

a) $A = \begin{bmatrix} 0 & 2 & -3 \\ 1 & 1 & 1 \\ 2 & -2 & 5 \end{bmatrix}$ b) $A = \begin{bmatrix} -5 & -6 & -3 \\ 14 & 11 & 7 \\ -10 & -2 & -4 \end{bmatrix}$ c) $A = \begin{bmatrix} -7 & 4 & 8 & -2 \\ -4 & 5 & 6 & -2 \\ -10 & 6 & 14 & -5 \\ -20 & 16 & 32 & -13 \end{bmatrix}$

18. If matrix A can be diagonalized (see exercise 17) $A = VDV^{-1}$, show that $A^k = VD^kV^{-1}$. Use this fact to compute A^5 for each matrix A in exercise 17. Use MATLAB to compare A^5 with V*D^5*inv(V).

19. If matrix A can be diagonalized (See exercise 17) $A = VDV^{-1}$, show that $e^A = Ve^DV^{-1}$. Hint: Use use exercise 18 and equation (10.22). Further, show that $e^{tA} = Ve^{tD}V^{-1}$ and use this result to calculate e^{tA} for each of the following matrices. Symbolic Toolbox owners should check their results with expm(t*A).

a) $A = \begin{bmatrix} -3 & 4 \\ -2 & 3 \end{bmatrix}$ b) $A = \begin{bmatrix} -8 & 9 \\ -6 & 7 \end{bmatrix}$ c) $A = \begin{bmatrix} 8 & 6 \\ -12 & -10 \end{bmatrix}$

20. The Cayley-Hamilton Theorem states that every matrix must satisfy its characteristic polynomial p_A; i.e., $p_A(A) = 0$. For example, the characteristic polynomial of

$$A = \begin{bmatrix} 5 & -5 & 3 \\ 2 & -1 & 2 \\ 0 & 1 & 2 \end{bmatrix}$$

is $p_A(\lambda) = \lambda^3 - 6\lambda^2 + 11\lambda - 6$. The following commands will verify that $p_A(A) = A^3 - 6A^2 + 11A - 6I = 0$.

```
>> A=[5 -5 3;2 -1 2;0 1 2];
>> A^3-6*A^2+11*A-6*eye(3)
```

Verify that each of the following matrices satisfies its characteristic polynomial. Note: You may also want to investigate the polyvalm command.

a) $\begin{bmatrix} 5 & 2 \\ -6 & -2 \end{bmatrix}$ b) $\begin{bmatrix} -1 & 1 \\ 6 & 4 \end{bmatrix}$ c) $\begin{bmatrix} -4 & 1 & 1 \\ 11 & 0 & -6 \\ -12 & 2 & 4 \end{bmatrix}$ d) $\begin{bmatrix} -1 & 1 & 0 \\ -1 & -2 & -1 \\ 3 & 2 & 2 \end{bmatrix}$

21. If A is 2×2 and has an eigenvalue c of multiplicity two, then the characteristic polynomial is $p_A(\lambda) = (\lambda - c)^2$. Because $p_A(A) = 0$ (See Exercise 20), $(A - cI)^2 = 0$. Each of the following 2×2 matrices has an eigenvalue c of multiplicity two. Find that eigenvalue c and verify that $(A - cI)^2 = 0$.

a) $\begin{bmatrix} 1 & 1 \\ -1 & 3 \end{bmatrix}$
b) $\begin{bmatrix} -3 & -1 \\ 1 & -1 \end{bmatrix}$
c) $\begin{bmatrix} 1 & -4 \\ 1 & 5 \end{bmatrix}$
d) $\begin{bmatrix} -6 & 1 \\ -1 & -4 \end{bmatrix}$

22. **Repeated Eigenvalue.** Use MATLAB to show that each of the following matrices have an eigenvalue of algebraic multiplicity two, but geometric multiplicity one. Use the identity $tA = \lambda t I + t(A - \lambda I)$, where λ is the eigenvalue of the matrix A, and the technique of Example 9 to simplify the exponential matrix e^{tA}.

a) $A = \begin{bmatrix} -4 & 1 \\ -1 & -2 \end{bmatrix}$
b) $A = \begin{bmatrix} -6 & -1 \\ 1 & -4 \end{bmatrix}$
c) $A = \begin{bmatrix} 3 & -1 \\ 1 & 1 \end{bmatrix}$
d) $A = \begin{bmatrix} 6 & -4 \\ 1 & 2 \end{bmatrix}$

23. **Symbolic Toolbox Users.** Use the Symbolic Toolbox command sequence `syms t`, `expm(t*sym(A))` to find each exponential matrix e^{tA} in Exercise 16.

24. **Repeated Eigenvalue.** Use MATLAB to show that each of the following systems have a repeated eigenvalue of algebraic multiplicity two, but geometric multiplicity one. Compute the exponential matrix. Use the exponential matrix and the technique of Example 9 to compute the solution of the given initial value problem.

a) $\mathbf{x}' = \begin{bmatrix} -3 & 1 \\ -4 & 1 \end{bmatrix} \mathbf{x}, \quad \mathbf{x}_0 = \begin{bmatrix} 1 \\ 1 \end{bmatrix}$
b) $\mathbf{x}' = \begin{bmatrix} 7 & -1 \\ 1 & 5 \end{bmatrix} \mathbf{x}, \quad \mathbf{x}_0 = \begin{bmatrix} -1 \\ 2 \end{bmatrix}$

25. **Symbolic Toolbox Users.** Use the Symbolic Toolbox to compute each solution in Exercise 24 via the exponential matrix. *Note: It is definitely challenging to get these solutions to match. Be persistent!*

26. The matrix

$$A = \begin{bmatrix} 3 & -1 & -2 \\ 14 & -5 & -4 \\ 11 & -2 & -7 \end{bmatrix}$$

has a repeated eigenvalue of algebraic multiplicity three. Use the technique of Example 9 to determine e^{tA}. Hint: See Exercises 20–21. Use the result to solve the initial value problem

$$x_1' = 3x_1 - x_2 - 2x_3,$$
$$x_2' = 14x_1 - 5x_2 - 4x_3,$$
$$x_3' = 11x_1 - 2x_2 - 7x_3,$$

where $x_1(0) = x_2(0) = x_3(0) = 1$.

27. Each of the following initial value problems yields a coefficient matrix with a repeated eigenvalue. Use the technique of Example 10 to find the solution.

a) $x' = 6x - 8y + 7z$
   $y' = 3x - 4y + 3z$
   $z' = -x + 2y - 2z$
   $x(0) = y(0) = z(0) = 1$

b) $x' = x + y - 3z$
   $y' = 10x - 2y - 10z$
   $z' = -2x + y$
   $x(0) = y(0) = z(0) = 1$

   3, -2, -2

c) $x_1' = 3x_1 + 4x_2 - 3x_3 - 12x_4$
   $x_2' = 2x_1 - 4x_2 + 6x_3 + 15x_4$
   $x_3' = -4x_1 - 5x_2 + 4x_3 + 15x_4$
   $x_4' = 2x_1 + x_2 - 2x_4$
   $x_1(0) = x_2(0) = x_3(0) = x_4(0) = 1$

Owners of the Symbolic Toolbox will want to check their answers with the `dsolve` command.

Al, Mult. - 1, 2

Geom. Mult - 1, 1

$V_1 =$ $V_2 =$

11. Advanced Use of PPLANE5

We introduced pplane5 in Chapter 6. In this chapter, we want to investigate some of the more advanced features of pplane5, such as drawing nullclines, finding and classifying equilibrium points, displaying linearizations, and drawing the separatrices that accompany "saddle" equilibrium points.

Recall that pplane5 can provide numerical solutions of planar, autonomous systems of the form

$$x' = f(x, y), \tag{11.1}$$
$$y' = g(x, y). \tag{11.2}$$

For purposes of discussion in this chapter, we assume three things:

- the vertical axis is the y-axis,
- the horizontal axis is the x-axis, and
- the independent variable is t.

Of course, as you saw in chapter 6, pplane5 allows you to change these defaults. We will not do so here.

Nullclines

Suppose that we take the right-hand side of equation (11.1) and set it equal to zero, as in $f(x, y) = 0$. The graph of $f(x, y) = 0$ can be drawn in the xy-phase plane. This graph is called a *nullcline*. More importantly, since $x' = f(x, y) = 0$, the derivative of x with respect to t is zero at each point (x, y) on this nullcline. Since $dx/dt = x' = 0$ for each point along this nullcline, there can be no right or left motion at points on this nullcline. Consequently, solution curves must be completely vertical at the instant they pass through this nullcline.

If we set the right-hand side of equation (11.2) equal to zero, then the plot of $g(x, y) = 0$ in the phase plane gives a second nullcline. More importantly, since $y' = g(x, y) = 0$, the derivative of y with respect to t is zero at each point (x, y) on this nullcline. Because $dy/dt = y' = 0$ for each point along this nullcline, there can be no up and down motion at points on this nullcline. Consequently, solution curves must be completely horizontal at the instant they pass through this nullcline.

Confused? Perhaps an example will clear the waters.

Example 1. *Consider the system*

$$\frac{dx}{dt} = 2x + 3y - 6, \tag{11.3}$$

$$\frac{dy}{dt} = 3x - 4y - 12. \tag{11.4}$$

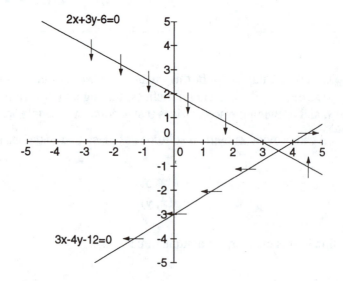

Figure 11.1. The nullclines of $dx/dt = 2x + 3y - 6$ and $dy/dt = 3x - 4y - 12$.

If we set the right-hand side of equation (11.3) equal to zero, we have $2x + 3y - 6 = 0$. The graph of $2x + 3y - 6 = 0$ is a nullcline, along which $dx/dt = 0$, and is pictured in Figure 11.1. Similarly, if we set the right-hand side of equation (11.4) equal to zero, then the equation of this nullcline is $3x - 4y - 12 = 0$, along which $dy/dt = 0$, and its graph is also pictured in Figure 11.1.

Furthermore, one can reason that $dx/dt = 2x + 3y - 6 > 0$ at all points (x, y) above the line $2x + 3y - 6 = 0$, and $dx/dt = 2x + 3y - 6 < 0$ at all points (x, y) below the line $2x + 3y - 6 = 0$. This implies that the direction field on the nullcline $3x - 4y - 12 = 0$ points to the right and left as indicated in Figure 11.1.

In a similar manner, $dy/dt = 3x - 4y - 12 < 0$ at all points (x, y) above the line $3x - 4y - 12 = 0$, and $dy/dt = 3x - 4y - 12 > 0$ at all points (x, y) below the line $3x - 4y - 12 = 0$. This implies that the direction field on the nullcline $2x + 3y - 6 = 0$ points up and down as indicated in Figure 11.1.

`Pplane5` completely automates this process. Begin by entering the equations $x' = 2x + 3y - 6$ and $y' = 3x - 4y - 12$ in the PPLANE5 Setup window. Set the display window so that $-5 \leq x \leq 5$ and $-5 \leq y \leq 5$, as shown in Figure 11.2.

Look carefully at Figure 11.2, and note that we have checked a different radio button than normal. Rather than selecting **Arrows** for the direction field, note that we have selected **Nullclines**.

Click the **Proceed** button to transfer information to the PPLANE5 Display window. Use your mouse and click the phase plane in several locations to start solution trajectories. You might not get exactly the same image shown in Figure 11.3, but you should get an image that demonstrates two very important points.

- Note that the nullclines are drawn in Figure 11.3, and arrows have been drawn on the nullclines indicating the direction that trajectories will follow when passing through the nullclines. Note that there are fewer arrows, but they do point in the same direction as those shown in Figure 11.1.
- Note that solution trajectories, at the instant they cross the nullclines, pass in a completely vertical

192

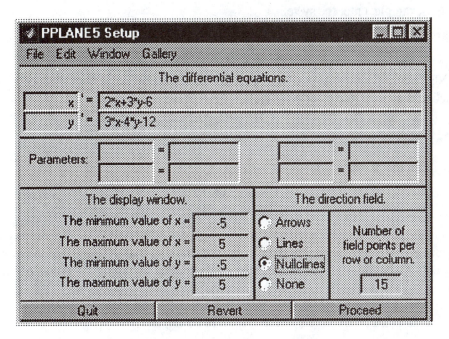

Figure 11.2. Setting up $x' = 2x + 3y - 6$ and $y' = 3x - 4y - 12$.

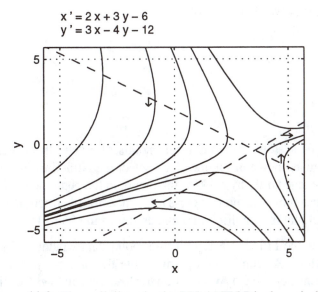

Figure 11.3. The nullclines in the PPLANE5 Display window

or horizontal orientation, depending on the nullcline.

Example 1 may leave you with the impression that the graph of a nullcline is always a line. This is not the case. For example, enter the system

$$x' = x^2 + y^2 - 4,$$
$$y' = x - y^2,$$

in the PPLANE5 Setup window, select **Nullclines** for the direction field, click **Proceed**, and you will note one nullcline is a circle, the other a parabola.

Equilibrium Points

Consider again the planar, autonomous system

$$x' = f(x, y), \tag{11.5}$$
$$y' = g(x, y), \tag{11.6}$$

A point (x, y) in the phase plane is called an *equilibrium point* if and only if two conditions are satisfied: $f(x, y) = 0$ and $g(x, y) = 0$. Note that these conditions require that the equilibrium point lie on *both* nullclines. Therefore, equilibrium points are most easily spotted by noting the points of intersection of nullclines.

Example 2. *Find all equilibrium points of the system in Example 1.*

Setting each of the right-hand sides of equations (11.3) and (11.4) equal to zero yields the following system of linear equations.

$$2x + 3y = 6,$$
$$3x - 4y = 12.$$

This linear system is easily solved in MATLAB.

```
>> A=[2 3;3 -4];
>> b=[6;12];
>> x=A\b
x =
    3.5294
   -0.3529
```

Pplane5 can also be used to find equilibrium points. Re-enter the system of Example 1 in the PPLANE5 Setup window as shown in Figure 11.1. Select the **Nullclines** for the direction field, click **Proceed**, then start a few trajectories by clicking the mouse in the phase plane. Select **Solutions→Find an equilibrium point**, then move the mouse "cross hairs" near the intersection of the nullclines and click the mouse. Pplane5 will plot the equilibrium point, as shown in Figure 11.4.

Pplane5 also opens the PPLANE5 Equilibrium point data window, which summarizes a number of important facts about the equilibrium point. Note the first entry, "There is a saddle point at (3.5294, −0.35294)," then click the **Go Away** button to close the window. Note how the solution trajectories in Figure 11.4 are somewhat "saddle-shaped," hence the name "saddle equilibrium point." Solution trajectories appear to approach the equilibrium point only to veer away at the last moment.

Example 3. *Find all equilibrium points of the system*

$$x' = x^2 + y^2 - 4, \tag{11.7}$$
$$y' = x - y^2. \tag{11.8}$$

194

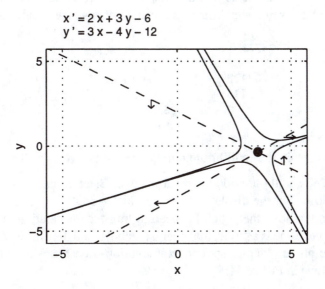

$$x' = 2x + 3y - 6$$
$$y' = 3x - 4y - 12$$

Figure 11.4. The equilibrium point is the intersection point of the nullclines.

Set the right-hand side of each differential equation equal to zero to obtain the following system of equations.

$$x^2 + y^2 - 4 = 0 \qquad (11.9)$$
$$x - y^2 = 0 \qquad (11.10)$$

This system is nonlinear, so matrix algebra is inappropriate here. You could use the substitution method, substituting equation (11.10) into equation (11.9), and follow with a lot of algebra to obtain two equilibrium points:

$$\left(\frac{-1 + \sqrt{17}}{2}, \frac{\sqrt{-2 + 2\sqrt{17}}}{2} \right) \quad \text{and} \quad \left(\frac{-1 + \sqrt{17}}{2}, -\frac{\sqrt{-2 + 2\sqrt{17}}}{2} \right).$$

Symbolic Toolbox Users. Owners of the Symbolic Toolbox can use the `solve` command to find the equilibrium points.

```
>> [x,y]=solve('x^2+y^2-4=0','x-y^2=0','x,y')
x =
[ -1/2+1/2*17^(1/2)]
[ -1/2+1/2*17^(1/2)]
[ -1/2-1/2*17^(1/2)]
[ -1/2-1/2*17^(1/2)]
y =
[  1/2*(-2+2*17^(1/2))^(1/2)]
[ -1/2*(-2+2*17^(1/2))^(1/2)]
[  1/2*(-2-2*17^(1/2))^(1/2)]
[ -1/2*(-2-2*17^(1/2))^(1/2)]
```

Careful inspection of this result reveals that the first two entries of vectors x and y contain symbolic results identical to our algebraic results. You might want to change the class of x and y to double.

```
>> S=[double(x),double(y)]
S =
    1.5616              1.2496
    1.5616             -1.2496
   -2.5616                   0 + 1.6005i
   -2.5616                   0 - 1.6005i
```

The first two rows of S contain the equilibrium points; the last two rows of S are complex, and ignored.

Of course, pplane5 should provide identical results. Enter equations (11.7) and (11.8) into the PPLANE5 Setup window. Set the display window so that $-5 \le x \le 5$ and $-5 \le y \le 5$. Select **Nullclines** for the direction field, then click **Proceed** to transfer information to the PPLANE5 Display window. Note that one nullcline is a circle, the other a parabola, and they have two points of intersection. Click the mouse in the phase plane to start several solution trajectories near each equilibrium point, similar to the image shown in Figure 11.5.

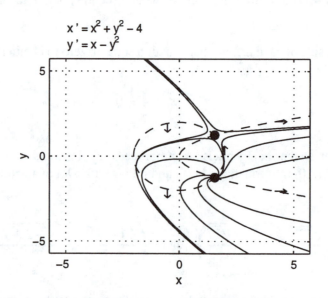

$$x' = x^2 + y^2 - 4$$
$$y' = x - y^2$$

Figure 11.5. Two equilibrium points: a saddle and a source.

Note that the higher of the two equilibrium points behaves like a "saddle," in a manner similar to the behavior of the equilibrium point in Example 2. Solution trajectories approach the equilibrium point, only to veer away at the last moment.

Select **Solutions→Find an equilibrium point** and click the mouse near the higher equilibrium point. Repeat the process with the lower equilibrium point. This will open the PPLANE5 Equilibrium point data window shown in Figure 11.6.

The information in this window indicates that the lower equilibrium point is a spiral source, located at approximately $(1.5616, -1.2496)$, numbers that are consistent with our earlier response from the

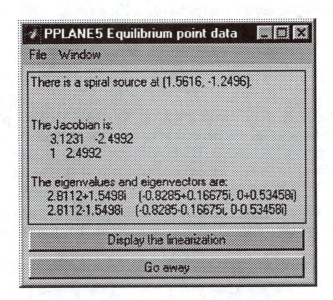

Figure 11.6. Description of the lower equilibrium point.

Symbolic Toolbox's `solve` command. Note that the solution trajectories in Figure 11.6 "spiral away" from the lower equilibrium point. Hence the name: spiral source.

Linear Systems

The system

$$x' = ax + by,$$
$$y' = cx + dy, \tag{11.11}$$

is planar, autonomous, and linear. The system can be written in the form

$$\begin{bmatrix} x \\ y \end{bmatrix}' = \begin{bmatrix} a & b \\ c & d \end{bmatrix} \begin{bmatrix} x \\ y \end{bmatrix},$$

which is equivalent to

$$\mathbf{x}' = A\mathbf{x}, \quad \text{where} \quad \mathbf{x} = \begin{bmatrix} x \\ y \end{bmatrix} \quad \text{and} \quad A = \begin{bmatrix} a & b \\ c & d \end{bmatrix}. \tag{11.12}$$

The equilibrium points are found by setting the right-hand side of each differential equation in system (11.11) equal to zero.

$$ax + by = 0$$
$$cx + dy = 0 \tag{11.13}$$

If the det $A \neq 0$, then system (11.13) has a unique solution, $(x, y) = (0, 0)$, which makes $(0, 0)$ an equilibrium point of system (11.11). Linear systems typically have a unique equilibrium point at $(0, 0)$, but strange things can happen when det $A = 0$. For example, the system $x' = x + 2y$, $y' = 2x + 4y$ has an entire line of equilibrium points. Experiment with this system in `pplane5`.

Pplane5 facilitates the study of linear systems.

197

Example 4. *Find and classify the equilibrium point of the linear system*

$$x' = -2x - 2y,$$
$$y' = 3x + y.$$

(11.14)

Select **Gallery→linear system** from the PPLANE5 Setup menu. At this point, note that most of the work has been done for you. All you have to do is adjust the parameters A, B, C, and D. Compare system (11.14) with the general linear system (11.11) and note that $a = -2$, $b = -2$, $c = 3$, and $d = 1$. Enter these parameters for A, B, C, and D in the PPLANE5 Setup window, then click **Proceed** to transfer information to the PPLANE5 Display window.

Select **Solutions→Keyboard input** and start a solution trajectory with initial condition $(0, 1)$. Note how this solution trajectory "spirals inward" toward the origin, as shown in Figure 11.7.

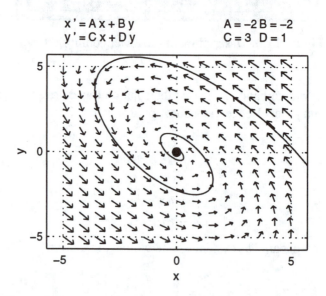

Figure 11.7. A spiral sink at the equilibrium point $(0, 0)$.

Select **Solutions→Find an equilibrium point** and click your mouse near the origin in the phase plane. Let's carefully examine the information provided in the PPLANE5 Equilibrium point data window shown in Figure 11.8.

There are a number of important entries in the PPLANE5 Equilibrium point data window.

- The first entry informs us that there is a spiral sink at the equilibrium point $(0,0)$. This certainly agrees with our analysis of the trajectory in Figure 11.7.
- Secondly, we are informed that the Jacobian matrix is

$$J = \begin{bmatrix} -2 & -2 \\ 3 & 1 \end{bmatrix}.$$

198

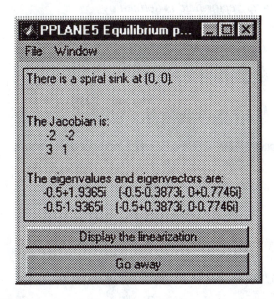

Figure 11.8. Description of the equilibrium point in Figure 11.7.

If system (11.14) is written in matrix form $\mathbf{x}' = A\mathbf{x}$, then the Jacobian J is identical to the system matrix A. This is always the case for linear systems.

- The eigenvalue-eigenvector pairs of the Jacobian are reported.
- Finally, there is a **Display the linearization** button, but this will provide no new information, as system (11.14) is already linear.

Nonlinear Systems

If a planar, autonomous system does not have the precise form of system (11.11), then the system is nonlinear. Nonlinear systems are much harder to analyze than linear systems. They rarely possess an analytical solution, so numerical solvers are a must when examining a system that is nonlinear.

The study of nonlinear systems presupposes that you have thoroughly investigated linear systems. There is a tacit assumption that you know all there is to know about linear systems and their equilibrium points. Why? Because the analysis of nonlinear systems is usually performed by approximating the behavior of a nonlinear system at an equilibrium point with a linear system, a technique called *linearization*.

If the term *linearization* sounds familiar, it should. In your first calculus class, you constantly analyzed the behavior of a function at a point by approximating the function with a first order Taylor's series, whose graph was called the *tangent line*. The graph of the tangent line was an excellent approximation of the function near the point of tangency. As you moved away from the point of tangency, the tangent line approximation was not so good.

We will use the same technique to analyze a nonlinear system at an equilibrium point. We will take a first order Taylor series approximation, called the Jacobian, and use this linearization to predict the behavior of the system near the equilibrium point. Again, as we move away from the equilibrium point, the approximation deteriorates.

Definition. *Suppose that a planar vector field is defined by*

$$F(x, y) = \begin{bmatrix} f(x, y) \\ g(x, y) \end{bmatrix}.$$

Then the Jacobian of **F** *is defined to be*

$$JF(x, y) = \begin{bmatrix} f_x(x, y) & f_y(x, y) \\ g_x(x, y) & g_y(x, y) \end{bmatrix},$$

where $f_x(x, y)$ *denotes the partial derivative of* f *with respect to* x, *evaluated at* (x, y), *etc.*

Example 5. *Consider the nonlinear system of differential equations*

$$\begin{aligned} x' &= 4x - 2x^2 - xy, \\ y' &= y - y^2 + 2xy. \end{aligned} \tag{11.15}$$

The right-hand side of system (11.15), in vector form, is written

$$F(x, y) = \begin{bmatrix} f(x, y) \\ g(x, y) \end{bmatrix} = \begin{bmatrix} 4x - 2x^2 - xy \\ y - y^2 + 2xy \end{bmatrix}.$$

Consequently, the Jacobian matrix is calculated as follows.

$$JF(x, y) = \begin{bmatrix} f_x(x, y) & f_y(x, y) \\ g_x(x, y) & g_y(x, y) \end{bmatrix} = \begin{bmatrix} 4 - 4x - y & -x \\ 2y & 1 - 2y + 2x \end{bmatrix} \tag{11.16}$$

Symbolic Toolbox Users. Owners of the Symbolic Toolbox can verify this last calculation of the Jacobian.

```
>> syms x y
>> F=[4*x-2*x^2-x*y;y-y^2+2*x*y]
F =
[ 4*x-2*x^2-x*y]
[   y-y^2+2*x*y]
>> JF=jacobian(F)
JF =
[   4-4*x-y,       -x]
[      2*y, 1-2*y+2*x]
```

You can also use the `subs` command to evaluate the Jacobian at a particular point.

```
>> subs(JF,{x,y},{3/4,5/2})
ans =
   -1.5000   -0.7500
    5.0000   -2.5000
```

Consequently,

$$JF(3/4, 5/2) = \begin{bmatrix} -1.5000 & -0.7500 \\ 5.0000 & -2.5000 \end{bmatrix}. \tag{11.17}$$

`Pplane5` can greatly simplify the analysis of nonlinear systems.

Example 6. *Consider the system*

$$x' = 2x(2 - x) - xy, \tag{11.18}$$

$$y' = y(1 - y) + 2xy. \tag{11.19}$$

Let x represent a rabbit population and y represent a fox population. Assume that both x and y are greater than or equal to zero.

A little algebra reveals that the system depicted by equations (11.18) and (11.19) is identical to that of system (11.15) of Example 5. We present the system here in factored form to facilitate discussion of the model.

If we deleted the xy terms from the right-hand sides of equation (11.18) and (11.19), both the rabbit and fox populations would grow according to the logistic model. The xy terms denote interaction between the two species. The presence of the term $-xy$ in equation (11.18) indicates that interaction between the species is harmful to the rabbit population, while the presence of the term $+2xy$ in equation (11.19) indicates that the interaction increases the growth rate of the fox population. Because the interaction among the species is beneficial to one but harmful to the other, this type of system is sometimes called a *predator-prey* model.

Enter equations (11.18) and (11.19) into the PPLANE5 Setup window. Set the display window so that $0 \leq x \leq 3$ and $0 \leq y \leq 3$. Select the **Arrows** radio button for the direction field and click **Proceed** to transfer information to the PPLANE5 Display window.

In the PPLANE5 Display window, select **Solutions→Plot nullclines** to overlay the nullclines on the vector field. Repeatedly select **Solutions→Find an equilibrium point** and use your mouse until you have found all four equilibrium points of the rabbit-fox system. The results are shown in Figure 11.9.

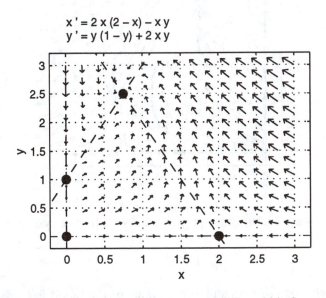

Figure 11.9. Phase plane and equilibrium points for rabbit-fox population.

Once you've found all of the equilibrium points, you can obtain a summary of the equilibrium points you found if you select **Solutions→List computed equilibrium points**. A summary of the computed

201

equilibrium points will be printed in the MATLAB workspace. You will have to switch to the command window to see the printed results[1].

```
(0.7500, 2.5000)    Spiral sink.
(0.0000, 1.0000)    Saddle point.
(0.0000, 0.0000)    Nodal source.
(2.0000, 0.0000)    Saddle point.
```

The point $(0, 0)$ is a nodal source. Solution trajectories starting near this point tend to emanate away from this point. Hence, the term "nodal source." Think of the trajectories as water flowing from a source in the ground. Click your mouse near the point $(0, 0)$ to witness this behavior.

The points $(0, 1)$ and $(2, 0)$ are saddle points. Click your mouse near these points and note that solutions tend toward these points, only to veer away at the last moment.

Finally, the equilibrium point at $(0.75, 2.5)$ is declared a spiral sink. Think of the drain in a bathtub, with water swirling around and into the drain. Click your mouse near the point $(0.75, 2.5)$ to witness this behavior.

If you've carefully followed the directions in the proceeding paragraphs, Figure 11.10 shows a possible phase portrait for the rabbit-fox system.

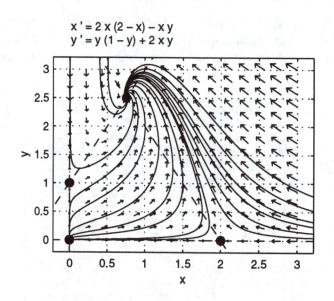

$$x' = 2\,x\,(2 - x) - x\,y$$
$$y' = y\,(1 - y) + 2\,x\,y$$

Figure 11.10. Phase portrait for the rabbit-fox system.

Linearization. Let revisit one of the equilibrium points of the rabbit-fox system, the spiral sink at $(0.75, 2.5)$. Select **Solutions**→**Find an equilibrium point** and click near the point $(0.75, 2.5)$ to re-open the PPLANE5 Equilibrium point data window, shown in Figure 11.11.

---

[1] Symbolic Toolbox owners will want to try
```
[x,y]=solve('2*x*(2-x)-x*y','y*(1-y)+2*x*y','x,y').
```

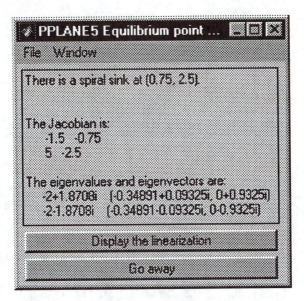

Figure 11.11. Data for the spiral sink at (0.75, 2.5).

The first line in the PPLANE5 Equilibrium point data window indicates a spiral sink at (0.75, 2.5). Then the Jacobian matrix is reported, evaluated at the equilibrium point (0.75, 2.5), and Jacobian's eigenvalue-eigenvector pairs are presented. Click the **Display the linearization** button to show the linear approximation of the rabbit-fox system shown in Figure 11.12.

The title heading in Figure 11.12 holds the key to the relationship between the Jacobian and the linear approximation. First, note that the equation of the linear approximation is given by

$$u' = Au + Bv$$
$$v' = Cu + Dv$$

More importantly, gaze to the right of the title header in the PPLANE5 Linearization window and note the values of A, B, C, and D. They are precisely the entries in the Jacobian matrix reported in the PPLANE5 Equilibrium point data window shown in Figure 11.11. So, you've now got the secret! If you want a linear approximation of a nonlinear system at an equilibrium point, compute the Jacobian at the equilibrium point, and form a linear system using the entries of the Jacobian matrix.

Actually, we are hiding a little mathematics from our readers. Note that the linearization in the PPLANE5 Linearization window has an equilibrium point at $(0, 0)$, not (0.75, 2.5). That's because linear systems always have an equilibrium point at $(0, 0)$. Consequently, when forming the linearization, we have to translate coordinate systems so that the origin of the linearization aligns itself with the equilibrium point (0.75, 2.5). That's why we're using u and v coordinates to define the linearization. They are the coordinates of the translated coordinate system.

Finally, what's the relationship between the eigenvalue-eigenvector pairs of the Jacobian and the type of equilibrium point shown in the PPLANE5 Equilibrium point data window? To answer this question, we place the linearization shown in the title of Figure 11.11 in matrix form.

$$\begin{bmatrix} u \\ v \end{bmatrix}' = \begin{bmatrix} -1.50 & -0.75 \\ 5.00 & -2.50 \end{bmatrix} \begin{bmatrix} u \\ v \end{bmatrix} \tag{11.20}$$

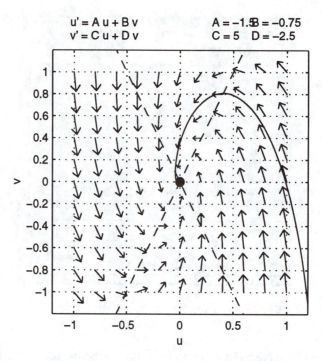

Figure 11.12. Linear approximation of the nonlinear rabbit-fox system at $(0.75, 2.5)$.

Using the technique demonstrated in Chapter 10, and the first eigenvalue-eigenvector pair in the PPLANE5 Equilibrium point data window, we can write the general solution of the linear system (11.20).

$$\begin{bmatrix} x \\ y \end{bmatrix} = c_1 e^{-2t} \begin{bmatrix} -0.34891 \cos 1.8708t - 0.09325 \sin 1.8708t \\ -0.9325 \sin 1.8708t \end{bmatrix}$$

$$+ c_2 e^{-2t} \begin{bmatrix} 0.09325 \cos 1.8708t - 0.34891 \sin 1.8708t \\ 0.9325 \cos 1.8708t \end{bmatrix} \qquad (11.21)$$

With experience, you will learn that it is not necessary to write the entire solution in order to determine the type of equilibrium point. The fact that the real part of the eigenvalue is -2 introduces a factor of e^{-2t}, accounting for the decay as time moves forward. Furthermore, the presence of complex numbers in the eigenvalue-eigenvector pair accounts for the rotation in the phase plane. Consequently, the equilibrium point of the linear system can be nothing other than a spiral sink.

The type of equilibrium point of the nonlinear system does not always match the type of equilibrium point of the linear approximation. There are exceptions, a number of which you will discover in the exercises at the end of the chapter. However, for the most part, the behavior near an equilibrium point of a nonlinear system will match the behavior near the equilibrium point of the linearization.

Zoom In Square Often, one would like to compare the phase portrait near an equilibrium point of a nonlinear system with the phase portrait of the linearization of the system near that same equilibrium point. The obvious approach is to find the linearization, plot a number of trajectories in this linearization window, then compare this result with the phase portrait near the equilibrium point in the PPLANE5 Display window.

However, the linearization is a good approximation of the nonlinear system only if you are "sufficiently close" to the equilibrium point. Usually, you will want to "zoom in" to make such comparisons.

But u ranges from -1 to 1 in Figure 11.12. Similarly, $-1 \leq v \leq 1$, giving the linearization window a "square look." If you try **Edit→Zoom in** within the PPLANE5 Display window, this can cause different unit lengths on the coordinate axes, giving a somewhat distorted or skewed-looking phase portrait, detrimental for comparison with the phase portrait in the linearization window. However, if you use **Edit→Zoom in square**, you can eliminate this distortion and get a good comparison of the phase portrait in the PPLANE5 Display window with that in the linearization window. Try it!

separatrices

Saddle equilibrium points have a special type of nearby solution trajectory called a separatrix. This is best demonstrated with an example.

Example 7. *Consider the nonlinear system*

$$x' = x(2 - x) - 2xy,$$
$$y' = y(2 - y) - 2xy.$$

(11.22)

Assume that x and y represent populations that are always non-negative.

If we remove the xy terms from system (11.22), then both populations would behave according to the logistic model. However, because of the minus signs, including the xy terms in this model is detrimental to the growth of *both* populations. This is an example of *competition*. Perhaps x and y represent sheep and cattle populations competing for the same grazing land.

Enter the equations in the PPLANE5 Setup window and set the display window so that $0 \leq x \leq 3$ and $0 \leq y \leq 3$. Select **Arrows** for the direction field and click **Proceed** to transfer information to the PPLANE5 Display window.

In the PPLANE5 Display window, select **Solutions→Plot nullclines** to overlay the nullclines on the direction field. Repeatedly select **Solutions→Find an equilibrium point** until you have found all four equilibrium points. Selecting **Solutions→List computed equilibrium points** will list the following summary in MATLAB's command window.

```
(0.0000, 2.0000)    Sink.
(0.0000, 0.0000)    Source.
(2.0000, 0.0000)    Sink.
(0.6667, 0.6667)    Saddle point.
```

Compute some solution trajectories by clicking various initial conditions in the phase plane. You will soon discover that some of the solutions tend to the sink at $(0, 2)$, spelling extinction for species x, while other trajectories tend to the sink at $(2, 0)$, indicating extinction of species y.

Select **Solutions→Plot stable and unstable orbits** and click near the saddle point at $(0.6667, 0.6667)$. Pplane5 will draw special types of trajectories called *separatrices*. Note that these separatrices "separate" solution trajectories that tend to the sink at $(0, 2)$ from the trajectories that tend to the sink at $(2, 0)$, as shown in Figure 11.13.

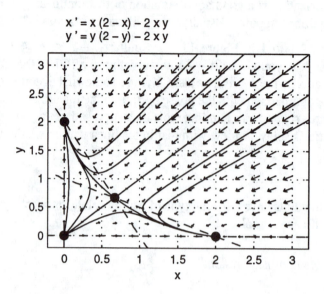

$$x' = x(2-x) - 2xy$$
$$y' = y(2-y) - 2xy$$

Figure 11.13. The separatrices divide the phase plane into basins of attraction.

Mathematicians will sometimes say that the separatrices divide the phase plane into "basins of attraction," meaning that initial conditions selected in one particular basin or region produce solutions that are attracted to one equilibrium point, while initial conditions in other regions create solutions that are attracted to different equilibrium points.

Improving Accuracy

In Chapter 7, in the section entitled "Improving Accuracy," we described how you can use options to improve the accuracy of MATLAB's solvers. The advice we gave in that section applies equally well to solutions created in `pplane5`.

Select **Option→Settings** to open the PPLANE5 Settings window. Here you can adjust the number of plot steps per computation step. This option is identical to the "Refine" option discussed in Chapter 7, "Improving Accuracy." You can also adjust the relative error tolerance. Again, we will refer readers to the "Improving Accuracy" section of Chapter 7. The description of relative error tolerance in Chapter 7 describes exactly how this option performs in `pplane5`.

Finally, there is a setting for determining the size of the calculation window. In truth, what you actually see in the PPLANE5 Display window is only a fraction of what's being computed. Having a computation window that's larger than what is actually displayed allows solutions to leave the display window and return. On slower machines, making the computation window smaller will speed the computation, but at the risk of losing re-entrant solutions. The default setting of 3.5 is sufficient for ordinary use of `pplane5`.

A Summary of Equilibrium Point Messages

`Pplane5` will recognize the types of many equilibrium points, but it is very conservative in its approach, and will often admit that it is unsure. Since `pplane5` examines an equilibrium point numerically, and because numerical algorithms always make small truncation errors, such a conservative approach is required. In particular, `pplane5` will only report information that is generic, i.e. information that will not change if the data on which it is based is changed by a small amount. Here is a complete list of the messages that can appear describing the equilibrium point, followed by an indication of what the message means.

```
There is a saddle point at ...
```

The eigenvalues are real and have different signs.

```
There is a nodal sink at ...
```

The eigenvalues are real, distinct, and negative. In some books this is called an *improper node*, a *stable improper node*, or a *stable node*.

```
There is a nodal source at ...
```

The eigenvalues are real, distinct, and positive. In some books this is called an *improper node*, an *unstable improper node*, or an *unstable node*.

```
There is a spiral sink at ...
```

The eigenvalues are complex conjugate and the real parts are negative. This is sometimes called a *spiral point*, a *stable spiral point*, or a *stable focus*.

```
There is a spiral source at ...
```

The eigenvalues are complex conjugate and the real parts are positive. This is sometimes called a *spiral point*, an *unstable spiral point*, or an *unstable focus*.

```
There is a spiral equilibrium point at ...
Its specific type has not been determined.
```

The eigenvalues are complex conjugate, with non-zero imaginary part. However, the real part is too close to 0 to decide whether this is a source, a sink, or a center. In particular, any true center (i.e. conjugate, purely imaginary eigenvalues) will be included in this category. A center is also called a *focus* in some books.

```
There is a source at ...
Its specific type has not been determined.
```

The eigenvalues have positive real part, but the eigenvalues are too close together to decide the exact type of the source. This could be a spiral source, a nodal source, or a degenerate source for which the Jacobian has equal eigenvalues. An equilibrium point of this type is unstable, but not much more can be said.

```
There is a sink at ...
Its specific type has not been determined.
```

The eigenvalues have negative real part, but the the eigenvalues are too close together to decide the exact type of the sink. This could be a spiral sink, a nodal sink, or a degenerate sink for which the

Jacobian has equal eigenvalues. An equilibrium point of this type is asymptotically stable, but not much more can be said.

```
There is an equilibrium point at ...

Its specific type has not been determined
```

At least one of the eigenvalues is so close to 0 that it is not possible to decide anything about the type of the equilibrium point. Closer analysis is required to describe the behaviour of the solutions.

It can be seen that `pplane5` will label an equilibrium point only when it is sure. Of course, more precise information about an equilibrium point can sometimes be garnered by looking at the Jacobian matrix. The Jacobian as provided by `pplane5` is a numerical approximation to the actual matrix of derivatives. Numerical approximations to derivatives are frequently inaccurate, so, especially in cases were there is doubt, the actual Jacobian should be computed and analyzed (either by hand or with the Symbolic Toolbox). The power of computers is still not sufficient to eliminate the need to do algebra.

However, looking at the Jacobian may not suffice, since this amounts to looking at the linearization of the system, and not the actual system. There are systems for which the linearization does not tell the whole story. Consequently, in the cases when `pplane5` cannot decide the precise type, the ultimate test of the type is what the solution curves actually do. Even then it is entirely possible that visualization will not provide the answer, and a rigorous analysis will be necessary.

Exercises

Nullclines and Equilibrium Points

1. Without the aid of technology, using only your algebra skills, sketch the nullclines and find the equilibrium point(s) of each of the following systems. Indicate the flow of the the vector field along each nullcline, similar to the flow shown in Figure 11.1. When you have completed this task, check your result with `pplane5`. Symbolic Toolbox owners may want to use the `solve` command to find the equilibrium point(s).

   a) $x' = x + 2y$
      $y' = 2x - y$

   b) $x' = x + 2y - 4$
      $y' = 2x - y - 2$

   c) $x' = 2x - y + 3$
      $y' = y - x^2$

   d) $x' = 2x + y$
      $y' = 4x + 2y$

Linear Systems.

Exercises 2–12 are designed to help classify all potential behavior of planar, autonomous, linear systems. It is essential that you master this material before attempting the analysis of non-linear systems. Be sure to turn **off** nullclines in these exercises.

Straight Line Solutions. If λ, $\mathbf{v}$ is an eigenvalue-eigenvector pair of matrix A, then we know that $\mathbf{x}(t) = e^{\lambda t}\mathbf{v}$ is a solution. Moreover, if λ and $\mathbf{v}$ are real, then this solution is a line in the phase plane. Because solutions of linear systems cannot cross in the phase plane, straight line solutions prevent any sort of rotation of solution trajectories about the equilibrium point at the origin.

2. In the exercises that follow, enter each system in `pplane5`, then find the eigenvalues and eigenvectors with the `eig` and `null` commands, as demonstrated in Example 4 of Chapter 10. You may find `format rat` helpful. Draw the straight line solutions. For example, if one eigenvector happens to be $\mathbf{v} = [1, -2]^T$, use the Keyboard input window to start straight line solutions at $(1, -2)$ and $(-1, 2)$. Perform a similar task for the other eigenvector. Finally, the straight line solutions in these exercises divide the phase plane into four regions. Use your mouse to start several solution trajectories in each region.

   a) $x' = 5x - y$
      $y' = -x + 5y$

   b) $x' = 6x - y$
      $y' = -3y$

   c) $x' = -x$
      $y' = 3x - 3y$

3. What condition on λ will ensure that the straight line solution $\mathbf{x}(t) = e^{\lambda t}\mathbf{v}$ moves toward the equilibrium point at the origin with the passage of time? What condition ensures that the straight line solution will move away from the equilibrium point with the passage of time?

4. If the eigenvalues and eigenvectors of a planar, autonomous, linear system are complex, then there are no straight line solutions. Use $[\mathbf{v},\mathbf{e}]=\mathtt{eig(A)}$ to demonstrate that the eigenvalues and eigenvectors of the system $x' = 2.9x + 2.0y$, $y' = -5.0x - 3.1y$ are complex, then use pplane5 to show that solution trajectories spiral in the phase plane.

5. The system $x' = x + y$, $y' = -x + 3y$ is a degenerate case, having only one straight line solution. Find the eigenvalues and eigenvectors of the system $x' = x + y$, $y' = -x + 3y$, then use pplane5 to depict the single straight line solution. Sketch several solution trajectories and note that they almost spiral, but the single straight line solution prevents full rotation about the equilibrium point at the origin. This degenerate case is a critical case, lying at a point that separates systems that spiral from those that do not.

Sources, Sinks, and Saddles. If a system has two distinct, real eigenvalues, then it will have two independent straight line solutions, and all solutions can be written as a linear combination of these two straight line solutions, as in $\mathbf{x}(t) = c_1 e^{\lambda_1 t}\mathbf{v}_1 + c_2 e^{\lambda_2 t}\mathbf{v}_2$.

6. **Nodal Sink.** If a system has two distinct negative eigenvalues, then both straight line solutions will decay to the origin with the passage of time. Consequently, all solutions will decay to the origin. Enter the system, $x' = -4x + y$, $y' = -2x - y$, in pplane5 and plot the straight line solutions. Plot several more solutions and note that they also decay to the origin. Select **Solutions→Find an equilibrium point**, find the equilibrium point at the origin, then read its classification from the PPLANE5 Equilibrium point data window.

7. **Nodal Source.** If a system has two distinct positive eigenvalues, then both straight line solutions will move away from the origin with the passage of time. Consequently, all solutions will move away from the origin. Enter the system, $x' = 5x - y$, $y' = 2x + 2y$, in pplane5 and plot the straight line solutions. Plot several more solutions and note that they also move away from the origin. Select **Solutions→Find an equilibrium point**, find the equilibrium point at the origin, then read its classification from the PPLANE5 Equilibrium point data window.

8. **Saddle.** If a system has one negative and one positive eigenvalue, then one straight line solution moves toward the origin and the other moves away. Consequently, general solutions (being linear combinations of straight line solutions) must do the same thing. Enter the system $x' = 9x - 14y$, $y' = 7x - 12y$, in pplane5 and plot the straight line solutions. Plot several more solutions and note that they move toward the origin only to move away at the last moment. Select **Solutions→Find an equilibrium point**, find the equilibrium point at the origin, then read its classification from the PPLANE5 Equilibrium point data window.

Spiral Source, Spiral Sink, Center. If the eigenvalues are complex, then a planar, autonomous, linear system has no straight line solutions. Consequently, trajectories are free to spiral.

9. **Spiral Source.** As we saw in equation (11.21), if the real part of a complex eigenvalue is negative, the equilibrium point will be a spiral sink. What happens if the real part of the complex eigenvalue is positive? Enter the system $x' = 2.1x + 5.0y$, $y' = -1.0x - 1.9y$ in pplane5 and note that solutions spiral outward with the passage of time. Select **Solutions→Find an equilibrium point**, find the equilibrium point at the origin, note the real part of the complex eigenvalue, then note the classification of the equilibrium point in the PPLANE5 Equilibrium point data window. If your computer is too fast to differentiate an outward spiral from an inward spiral, you might try **Options→Solution direction→Forward**.

10. **Center.** If the real part of the complex eigenvalue is zero, there can be no growth or decay of solution trajectories. Enter the system $x' = -x + y$, $y' = -2x + y$, and note that pplane5 reports a nearly closed orbit in its message window. Note that Pplane5 has difficulty with this type of equilibrium point, commonly called a *center*. It should report that the trajectories are purely periodic. This happens because this system is a critical case, separating linear systems with a spiral sink from systems with a spiral source. The slightest numerical error, even an error of 1×10^{-16}, makes the real part of the complex eigenvalue non-zero, indicating a spiral source or sink.

Summarizing Your Findings

11. Prepare a table, then use your results from Exercises 6–10 to list conditions on the eigenvalue that predict when the equilibrium point at the origin is a nodal source, nodal sink, saddle, spiral source, spiral sink, and

center.

The Trace-Determinant Plane. It is a nice exercise to classify linear systems based on their position in the trace-determinant plane.

12. Consider the matrix

$$A = \begin{bmatrix} a & b \\ c & d \end{bmatrix}.$$

a) Show that the characteristic polynomial of the matrix A is $p(\lambda) = \lambda^2 - T\lambda + D$, where T is the trace of matrix A, defined as $T = a + d$, and D is the determinant of matrix A, defined as $D = \det(A) = ad - bc$.

b) We know that the characteristic polynomial factors as $p(\lambda) = (\lambda - \lambda_1)(\lambda - \lambda_2)$, where λ_1 and λ_2 are the eigenvalues. Use this and the result of part (a) to show that the product of the eigenvalues is equal to the determinant of matrix A. *Note: This is a useful fact. For example, if the determinant is negative, then you must have one positive and one negative eigenvalue, indicating a saddle equilibrium point.* Also, show that the sum of the eigenvalues equals the trace of matrix A.

c) Show that the eigenvalues of matrix A are given by the formula

$$\lambda = \frac{T \pm \sqrt{T^2 - 4D}}{2}.$$

Note that there are three possible scenarios. If $T^2 - 4D < 0$, then there are two complex eigenvalues. If $T^2 - 4D > 0$, there are two real eigenvalues. Finally, if $T^2 - 4D = 0$, then there is one repeated eigenvalue of algebraic multiplicity two.

d) Draw a pair of axes on a piece of posterboard. Label the vertical axis D and the horizontal axis T. Sketch the graph of $T^2 - 4D = 0$ on your posterboard. The axes and the parabola defined by $T^2 - 4D = 0$ divide the trace-determinant plane into six distinct regions, as shown in Figure 11.14.

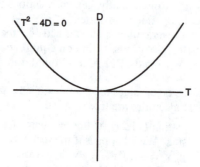

Figure 11.14. The trace-determinant plane.

e) You can classify any matrix A by its location in the trace-determinant plane. For example, if

$$A = \begin{bmatrix} 1 & 2 \\ -3 & 2 \end{bmatrix},$$

then $T = 3$ and $D = 8$, so the point (T, D) is located in the first quadrant. Furthermore, $(3)^2 - 4(8) < 0$, placing the point $(3, 8)$ above the parabola $T^2 - 4D = 0$. Finally, if you substitute $T = 3$ and $D = 8$ into the formula $\lambda = (T \pm \sqrt{T^2 - 4D})/2$, then you get eigenvalues that are complex with a positive real part, making the equilibrium point of the system $\mathbf{x}' = A\mathbf{x}$ a spiral source. Use `pplane5` to generate a phase portrait of this particular system and attach the printout to the posterboard at the point $(3, 8)$.

210

f) Linear systems possess approximately fourteen distinctive phase portraits. Attach a printout of each case at its appropriate point (T, D) in the trace-determinant plane on your posterboard. *Hint: There are degenerate cases on the axes and the parabola. For example, you can find degenerate cases on the parabola in the first quadrant that separate nodal sources from spiral sources. There are also a number of interesting degenerate cases at the origin of the trace-determinant plane. One final note: We have intentionally used the word "approximately" when implying the number of distinctive cases so as to spur argument amongst our readers when working on this activity. There may be more, or less. What do you think?*

Nonlinear Systems The essential analytical tool is linearization. That is, we analyze the behavior of a nonlinear system at an equilibrium point with a linear approximation.

13. **The Jacobian.** Consider the nonlinear system

$$x' = x(1 - x) - xy,$$
$$y' = y(2 - y) + 2xy.$$

Show that $(-1/3, 4/3)$ is an equilibrium point of the system.

a) Without the use of technology, calculate the Jacobian of the system at the equilibrium point $(-1/3, 4/3)$. What is the equation of the linearization at this equilibrium point? Use `[v,e]=eig(J)` to find the eigenvalues and eigenvectors of this Jacobian.

b) Enter the system in `pplane5`. Find the equilibrium point at $(-1/3, 4/3)$. Does the data in the Equilibrium point data window agree with your findings in part (a)? *Note that the eigenvalues of the Jacobian predict classification of the equilibrium point. In this case, the point $(-1/3, 4/3)$ is a saddle because the eigenvalues are real and opposite in sign.*

c) Display the linearization. Does the equation of the linearization agree with your findings in part (a)?

14. **Linearization.** Consider the system

$$x' = x^2 + xy - 2x,$$
$$y' = 3xy + y^2 - 3y.$$

Enter the system in `pplane5`.

a) Find the equilibrium point at $(1/2, 3/2)$. Display the linearization, plot the stable and unstable orbits and several solution trajectories to create a phase portrait of the linearization.

b) Return to the PPLANE5 Display window. Select **Edit→Zoom in square**, place the mouse at the equilibrium point $(1/2, 3/2)$, then drag and release a small zoom box. On a PC, you can depress the shift key, then drag with the left mouse button to achieve the same effect. Repeat several times until you have zoomed sufficiently close to the equilibrium point. Plot the stable and unstable orbits and several trajectories to create a phase portrait of the nonlinear system near $(1/2, 3/2)$. Compare this phase portrait with the phase portrait in the linearization window.

15. **When Linearization Fails.** Most of the time, the linearization accurately predicts the behavior of a nonlinear system at an equilibrium point. There are exceptions, most notably when the the matrix A has purely imaginary eigenvalues, or when one of the eigenvalues is zero. For example, consider the system

$$x' = -y + x(x^2 + y^2),$$
$$y' = x + y(x^2 + y^2).$$

Enter the system in `pplane5`, set the display window so that $-1 \le x \le 1$ and $-1 \le y \le 1$, select **Arrows** for the direction field, then click **Proceed**. In the PPLANE5 Display window, select **Solutions→Plot nullclines** to overlay the nullclines on the vector field and note the presence of an equilibrium point at $(0, 0)$.

a) Use Keyboard input to start a solution trajectory at $(0.5, 0)$ and note that the origin behaves as a spiral source. Well, maybe. It's strange that the solution trajectory stops short of decaying to the origin.

b) Select **Solutions→Find an equilibrium point** and find the equilibrium point at $(0, 0)$. Note that the eigenvalues of the Jacobian are purely imaginary, indicating that the linearization has a *center*, not a spiral source, at the origin. Display the linearization and draw some solution trajectories.

211

c) The difficulty in this analysis is the stopping criteria built into `pplane5`'s solver. When the solver doesn't detect much change over a significant length of time, it simply stops and reports a reason for the stoppage in the message window, in this case the detection of a nearly closed orbit. We'll have to work around this behavior.

In the PPLANE5 Setup window, change the display window setting so that $-0.1 \leq x \leq 0.1$ and $-0.1 \leq y \leq 0.1$. Click **Proceed** to return to the PPLANE5 Display window, erase all solutions, then use Keyboard input to start a solution trajectory at $(0.08, 0)$. Press the **Stop** button (if necessary) to stop the forward solution, then wait until the backward solution stops itself with the detection of a nearly closed orbit. Now, use Keyboard input to start a solution trajectory inside this orbit at $(0.02, 0)$. This orbit stops pretty quickly with the detection of a nearly closed orbit. Use Keyboard input to start the solution trajectory at $(0.02, 0)$ again, but this time **Specify the computation interval** so that $-1000 \leq t \leq 0$. Note that this forces the backward trajectory to spiral further inward, lending credence to the existence of a spiral source at $(0, 0)$.

16. Consider the differential system

$$\mathbf{x}' = \mathbf{A}\mathbf{x},$$

where **A** is one of the following matrices.

a) $\begin{bmatrix} 1 & 1 \\ -18 & 10 \end{bmatrix}$ b) $\begin{bmatrix} 6 & -1 \\ 0 & -3 \end{bmatrix}$ c) $\begin{bmatrix} -1 & 0 \\ 3 & -3 \end{bmatrix}$ d) $\begin{bmatrix} 6 & 1 \\ -18 & 0 \end{bmatrix}$ e) $\begin{bmatrix} 9 & -1 \\ 9 & 3 \end{bmatrix}$

f) $\begin{bmatrix} 9 & 1 \\ 9 & 3 \end{bmatrix}$ g) $\begin{bmatrix} 7 & 4 \\ -10 & -5 \end{bmatrix}$ h) $\begin{bmatrix} -1 & 0 \\ 0 & -1 \end{bmatrix}$ i) $\begin{bmatrix} 6 & -4 \\ 18 & -6 \end{bmatrix}$ j) $\begin{bmatrix} 2 & -2 \\ 4 & -4 \end{bmatrix}$

i) In each of these cases find the type of the equilibrium point at the origin. Do this without the aid of any technology or `pplane5`. You may, of course, check your answer using `pplane`.

ii) For each of these cases use `pplane5` to plot several solution curves — enough to fully illustrate the behavior of the system. You will find it convenient to use the "linear system" choice from the **Gallery** menu. If the eigenvalues are real, use the "Keyboard input" option to include straight line solutions starting at ± 10 times the eigenvectors. If the equilibrium point is a saddle point, compute and plot the separatrices.

17. For the linear systems in Exercise 16, consider the effect of a non-linear perturbation which leaves the Jacobian at $(0, 0)$ unchanged. The theory predicts that, in (maybe very small) neighborhoods of equilibrium points where the eigenvalues of the Jacobian are non-zero and not equal, a non-linear system will act very much like the linear system associated with the Jacobian. In this exercise we will use `pplane5` to verify this. To be precise, instead of 16a) consider the perturbed system

$$x' = x + y - xy,$$
$$y' = -18x + 10y - x^2 - y^2.$$

i) Show that the Jacobian of the perturbed system at the origin is the same as that of the unperturbed system.

ii) Starting with the display rectangle $-5 \leq x \leq 5$, $-5 \leq y \leq 5$, zoom in to smaller squares centered at $(0, 0)$ until the solution curves look like those of the linear system in 16a).

Repeat the experiment for the systems in 16b), 16c), and 16d) using the same or other quadratic or cubic terms as perturbations. Be sure that the perturbation vanishes to second order at the origin.

18. In contrast to the previous problem, consider the system

$$x' = y + ax^3,$$
$$y' = -x,$$

for the three values 0, 10 and -10 of the parameter a.

a) Show that all three systems have the same Jacobian matrix at the origin. What type of equilibrium point at $(0, 0)$ is predicted by the eigenvalues of the Jacobian?

b) Use `pplane5` to find evidence that will enable you to make a conjecture as to the type of the equilibrium point at $(0, 0)$ in each of the three cases.

c) Consider the function $h(x, y) = x^2 + y^2$. In each of the three cases, restrict h to a solution curve and differentiate the result with respect to t (Recall: $\frac{dh}{dt} = \frac{\partial h}{\partial x}\frac{dx}{dt} + \frac{\partial h}{\partial y}\frac{dy}{dt}$.). Can you use the result to verify the conjecture you made in part b)? *Hint: Note that $h(x, y)$ measures the distance between the origin and (x, y).*

d) Does the Jacobian predict the behavior of the non-linear systems in this case?

19. Consider the system
$$x' = ax + y,$$
$$y' = -x - 2y.$$

Describe what happens to the equilibrium points as a varies among the five choices -0.1, -0.01, 0, 0.01, and 0.1. Provide plots of each with enough solution curves to illustrate the behavior of the system. If the eigenvalues are real, use the **Keyboard input** option to include straight line solutions starting at ± 1 times the eigenvectors.

20. Consider the damped pendulum when the damping constant is D=0.5 versus when D=2.5.

a) Select **Gallery→pendulum** and use `pplane5` to examine the behavior of the system in each of these cases.

b) You will notice that the type of the equilibrium point at the origin is different for D=0.5 and D=2.5. Analyze the Jacobian matrix and find out at which value of D the type changes.

21. This is another example of how a system can change as a parameter is varied. Consider the system
$$x' = ax + y - x(x^2 + y^2),$$
$$y' = -x + ay - y(x^2 + y^2),$$

for $a = 0$, and $a = \pm 0.1$. Use the display rectangle $-1 \leq x \leq 1$, and $-1 \leq y \leq 1$, and plot enough solutions in each case to describe behavior of the system. Describe what happens as the parameter a varies from negative values to positive values. (This is an example of a *Hopf bifurcation*.) *Hint: Look for something completely new, something other than an equilibrium* **point**.

22. This problem involves the default system in `pplane5`, i.e., the planar system that appears when `pplane5` is first started.

a) Find and plot all interesting features, i.e., equilibrium points, separatrices, limit cycles, and any other features that appeal to you. Make a list of such to turn in with the plot.

b) Find the linearization of the system at $(0, 0)$. Enter that into `pplane5` and plot a few orbits.

c) Go back to the default system. According to theory, the linearization should approximate the original system in a small enough neighborhood of $(0, 0)$. Use the **Zoom in** option to find a small enough rectangle containing $(0, 0)$, where the behavior of the system is closely approximated by the linear system you analyzed in part b).

d) Redo parts b) and c) for the other equilibrium points of the default system.

23. The default system in `pplane5` exhibits two bifurcations involving the saddle point at the origin and the equilibrium point near $(3/2, -3/2)$. To be precise we are looking at the system
$$x' = 2x - y + 3(x^2 - y^2) + 2xy,$$
$$y' = x - 3y - 3(x^2 - y^2) + Axy,$$

as the parameter A varies.

a) Show that between $A = 2.5$ and $A = 3$, the equilibrium point changes from a sink to a source and spawns a limit cycle, i.e., a Hopf bifurcation occurs.

213

b) Show that between $A = 3$ and $A = 3.2$ the limit cycle disappears. At some point in this transition, the limit cycle becomes a homoclinic orbit for the saddle point at the origin. A *homoclinic orbit* is one that originates as an unstable orbit for saddle point at $t = -\infty$, and then becomes a stable orbit for the same saddle point as $t \to +\infty$. It is extremely difficult to find the value of A for which this occurs, so do not try too hard. The process illustrated here is called a *homoclinic bifurcation.*

c) The default system has four quadratic terms. Show that if any of the coefficients of these terms is altered up or down by as little as 0.5, a bifurcation takes place. Show that there is a Hopf bifurcation in one direction, and a homoclinic bifurcation in the other.

24. If $\mathbf{x}_0$ is a sink for the system $\mathbf{x}' = \mathbf{F}(\mathbf{x})$, then the *basin of attraction* for $\mathbf{x}_0$ is the set of points $\mathbf{y} = [y_1, y_2]^T$ which have the property that the solution curve through $\mathbf{y}$ approaches $\mathbf{x}_0$ as $t \to +\infty$. Consider the damped pendulum with damping constant $a = 0.5$. Find the basin of attraction for $[0, 0]^T$. You should indicate this region on a printed output of `pplane5`. *Hint: It will be extremely helpful to plot the stable and unstable orbits from a couple of saddle points.*

25. The system of differential equations:

$$x' = \mu x - y - x^3,$$
$$y' = x,$$

is called the *van der Pol system.* It arises in the study of non-linear semiconductor circuits, where y represents a voltage and x the current. It is in the **Gallery** menu.

a) Find the equilibrium points for the system. Use `pplane5` only to check your computations.

b) For various values of μ in the range $0 < \mu < 5$, find the equilibrium points, and find the type of each, i.e, is it a nodal sink, a saddle point, ...? You should find that there are at least two cases depending on the value of μ. Don't worry too much about non-generic cases. Use `pplane5` only to check your computations.

c) Use `pplane5` to illustrate the behavior of solutions to the system in each of the cases found in b). Plot enough solutions to illustrate the phenomena you discover. Be sure to start some orbits very close to $(0, 0)$, and some near the edge of the display window. Put arrows on the solution curves (by hand after you have printed them out) to indicate the direction of the motion. (The display window $(-5, 5, -5, 5)$ will allow you to see the interesting phenomena.)

d) For $\mu = 1$ plot the solutions to the system with initial conditions $x(0) = 0$, and $y(0) = 0.2$. Plot both components of the solution versus t. Describe what happens to the solution curves as $t \to \infty$.

26. *Duffing's equation* is

$$mx'' + cx' + kx + lx^3 = F(t).$$

When $k > 0$ this equation models a vibrating spring, which could be soft ($l < 0$) or hard ($l > 0$) (See Student Project #1 in Chapter 7). When $k < 0$ the equation arises as a model of the motion of a long thin elastic steel beam that has its top end embedded in a pulsating frame (the $F(t)$ term), and its lower end hanging just above two magnets which are slightly displaced from what would be the equilibrium position. We will be looking at the unforced case (i.e. $F(t) = 0$), with $m = 1$.

a) This is the case of a hard spring with $k = 16$, and $l = 4$. Use `pplane5` to plot the phase planes of some solutions with the damping constant $c = 0, 1,$ and 4. In particular, find all equilibrium points.

b) Do the same for the soft spring with $k = 16$ and $l = -4$. Now there will be a pair of saddle points. Find them and plot the stable/unstable orbits.

c) Now consider the case when $k = -1$, and $l = 1$. For each of the cases $c = 0$, $c = 0.2$, and $c = 1$, use `pplane5` to analyze the system. In particular, find all equilibrium points and determine their types. Plot stable/unstable orbits where appropriate, and plot typical orbits.

d) With $c = 0.2$ in part c), there are two sinks. Determine the basins of attraction of each of these sinks. Indicate these regions on a print out of the phase plane.

27. A wide variety of phenomena can occur when an equilibrium point is completely degenerate, i.e., when the

Jacobian is the zero matrix. We will look at just one. Consider the system

$$x' = xy,$$
$$y' = x^2 - y^2.$$

a) Show that the Jacobian at the origin is the zero matrix.

b) Plot the solutions through the six points $(0, \pm 1)$, and $(\pm\sqrt{2}, \pm 1)$. Plot additional solutions of your choice.

c) Compare what you see with the behavior of solutions near a saddle point.

Index

Items that appear in typewriter font, such as `dfield`, refer to MATLAB commands or variables.